LAPLACE TRANSFORM SOLUTION
OF DIFFERENTIAL EQUATIONS

A PROGRAMMED TEXT

LAPLACE TRANSFORM SOLUTION
OF DIFFERENTIAL EQUATIONS

ROBERT D. STRUM / JOHN R. WARD

Associate Professors of Electrical Engineering
Naval Postgraduate School, Monterey, California

PRENTICE-HALL, INC., ENGLEWOOD CLIFFS, N.J.

Library of Congress Catalog Card Number 68-11403

Printed in the United States of America

Prentice-Hall International, Inc., *London*
Prentice-Hall of Australia, Pty. Ltd., *Sydney*
Prentice-Hall of Canada, Ltd., *Toronto*
Prentice-Hall of India Private Ltd., *New Delhi*
Prentice-Hall of Japan, Inc., *Tokyo*

Current printing (last digit):
19   18   17   16   15   14   13   12

This program consists of an introduction to the Laplace transform solution of ordinary linear differential equations. As such it has been designed to lay down a firm foundation for the study of linear-system dynamics. To provide direction, and hence motivation, example problems have been drawn from various fields of engineering.

Essential prerequisites are a course in the calculus, the ability to solve simultaneous algebraic equations by determinants, and a knowledge of complex numbers through the Euler relation. A prior introduction to the *formulation* of system differential equations can be expected to contribute additional motivation, but is not otherwise prerequisite.

The program is intended for both class and individual self-study. It has been used successfully in a variety of situations—ranging from an introductory electrical fields and circuits course (with no prior course in differential equations) to an early graduate course in linear-system analysis. Individual undergraduate and graduate students have found the program helpful both as a first introduction to the Laplace transform and as a review of a long-forgotten technique.

PROGRAMMED INSTRUCTION

It is the authors' belief that programmed instruction has much to offer both instructor and student. This is not to say that whole courses should be programmed, but that the programming of certain key concepts and techniques can be mutually beneficial. To the student a program need not be impersonal, and it can in fact offer a flexibility and a degree of interaction which approach private tutoring. The instructor is provided with a virtual guarantee that all of his students who complete the program will demonstrate mastery of the material treated. He can then proceed to elaboration and application appropriate to the needs of his curriculum with the knowledge that he is building on strength.

These benefits stem from the basic characteristics of programmed instruction:

1. The student is required to participate actively throughout. He "learns by doing."

2. The instruction is self-paced. A student progresses in carefully sequenced steps; he cannot be forced ahead of his comprehension; he cannot start a new topic before he has developed a practiced competence in its prerequisites.

3. The student is provided with a continuous measure of his performance, and a source of assistance is always at hand.

This suggests, intuitively at least, why programmed instruction can be successful—it is geared to the needs and characteristics of the learner.*

At least as important is the fact that the preparation of a program lends itself to the engineering processes of specification, design, test, re-design, and evaluation.† A program is designed to meet given specifications or objectives on the basis of a knowledge of the learning process; it is tested in fine detail in the actual learning environment; shortcomings are corrected; and finally the program is evaluated by comparing student test (exam) results with the original specifications.

TEACHING FORMAT

A glance at any page of this program will reveal a characteristic sequencing of "question" frames on the left, with the corresponding "answers" on the right. The student responds to each frame, checks his answer, and goes to the next frame. A mask is used to obscure the answers until the corresponding responses have been made.

It is worth noting that one great advantage of programs in book format is that they are portable and always available to the student. This is not generally true of teaching machines, carrels, or instructional computer terminals. Further, the book can be retained by the student for later reference, and experience has shown that this is appreciated.

*For a more complete discussion, see: B. F. Skinner, "Teaching Machines," *Scientific American*, November 1961. (Note that a book can be a "teaching machine.")

†Norman Balabanian, "The Educational Engineering Called Programmed Instruction," *IEEE Transactions on Education*, June 1966.

ACKNOWLEDGMENTS

Our early struggles with this technology were in part supported by funds from the Naval Postgraduate School, the Ford Foundation, and other sponsors of the ASEE Programmed Learning Project of 1965–66.* The authors were particularly grateful for the opportunity to participate in this Project and for the impetus provided by the 1965 Summer Workshop, which was organized and led by Norman Balabanian and James Holland. Thanks are also due A. A. Root, Project editor, for his enthusiasm and critical direction.

For the technical content of the program the authors can only say that they are indebted to many former teachers, students, colleagues, and textbook authors.

Finally, our thanks are due to the students who worked through the program to provide us with evidence of its failings. We trust that through their difficulties we have progressed to an effective program.

ROBERT D. STRUM / JOHN R. WARD

*Norman Balabanian, "The Programmed Learning Project: A Progress Report," *Journal of Engineering Education*, April 1966.

TO THE INSTRUCTOR

The basic characteristics and *raison d'être* of programmed instruction have been outlined in the Preface.* Here the technical aspects of this particular program will be described.

The Table of Contents defines the general scope and organization of the program, while a perusal of the frames referenced therein will establish rapport with the authors' detailed approach to the subject matter. Briefly, attention has been concentrated on the most vital elements of the Laplace transform method, leaving the instructor with an open-ended opportunity to extend the treatment as he sees fit. Emphasis has been placed on the *technique* of equation solving—questions of convergence, existence, and uniqueness are avoided, and inverse transformation by contour integration is not mentioned. After much discussion it was decided that the initial time should be regarded as $t = 0$, rather than as $t = 0+$ or $t = 0-$. In passing, the definite integral has been used throughout. For example, the voltage on a capacitor is written

$$v(t) = \frac{1}{C} \int_0^t i(t)\,dt + v(0) \qquad \text{rather than as} \qquad v(t) = \frac{1}{C} \int i(t)\,dt$$

*For additional background see:

1. Robert Glasser (Ed.), "*Teaching Machines and Programmed Learning, II, Data & Directions,*" National Educational Association, 1965.
2. John P. DeCecco (Ed.), "*Educational Technology,*" Holt, Rinehart & Winston, N.Y., 1965.
3. E. R. Hilgard and G. H. Bower, "*Theories of Learning,*" 3rd Ed., Appleton-Century-Crofts, N.Y., 1966.

STUDENT PREREQUISITES

At the start of the program are 17 problems designed to test the student's background. No student having difficulty with any part of this test should be permitted to continue until the deficiency has been rectified.

PROGRAM OBJECTIVES

A careful and complete specification of objectives is a characteristic of programmed instruction.* In this sense the program has been designed to develop in the student the ability to solve problems such as those in Sections 19 through 22 and in the sample examination.

PROGRAM EVALUATION

Some three hundred students have used the program in a variety of classes taught by a total of nine instructors during the last year of test and evaluation. Program revisions have been based on analyses of student responses (and comments) in the program itself, on test performances (using the exam now incorporated in the text), and on formal observations of student attitudes. The outcome is a program in which the student error rate will normally be well under 10% (excluding errors of algebraic carelessness). Further, at least 80% of a class can be expected to score 90% or more in a test such as the one at the end of the program and the remaining 20% should score over 80%. This assumes, of course, that the students have performed acceptably in the test of prerequisites and have made an honest attempt to work through the program. There should be no difficulty on the latter score—the program has been enthusiastically received.

SUGGESTIONS FOR CLASS USE

A safe procedure is simply to ask the students to work through the program. No special comments or instructions are needed, and it has been found unnecessary to "sell" the idea of programmed instruction—the students usually return to sell the technique to the instructor!

The authors suggest that class meetings be suspended for about a week and that there should be no homework or exams on other aspects of the course during this time. Since the program includes both "lecture" and "homework" material, these conditions are not unreasonable and they will ensure that the student is free to concentrate on the program.

*Robert F. Mager, "*Preparing Instructional Objectives*," Fearon Publishers, Palo Alto, California, 1962.

A final test is recommended as soon as possible after the program has been completed. This test may be along the lines of the sample at the end of the program or an abbreviated version thereof. In every case, care *must* be taken to test *only* those skills developed by the program. Questions seeking, perhaps, to "separate the men from the boys," may easily destroy the whole spirit of programmed instruction. The students are told implicitly that they will *not* be asked to recall any result, identity, comment, or technique which they have not *practiced* in the program.

Finally, the students should be made aware of the following:

1. The class period by which the program is to be completed;

2. The times when the instructor will be available to answer questions; and

3. The class period(s) during which they will be required to work the final test.

xiii

LAPLACE TRANSFORM SOLUTION
OF DIFFERENTIAL EQUATIONS

```
┌─────────────────────┐
│                     │
│   THE PROGRAM       │
│   STARTS HERE       │
│                     │
└─────────────────────┘
```

TO THE STUDENT

Welcome! This is a book which has been designed for you. In fact, we know you can't avoid learning all that is expected of you *if* you play the game according to the rules. It will be hard work, but we will be surprised if you do not find it to be fun. Good luck!

We will start out with a quick test of your prerequisites. This should not take you longer than half an hour or so, and it will help you get used to working with a programmed text. *Don't skip it, please*.

First, TEAR OUT THE MASK (which is inside the front cover) AND SLIP IT UNDER THIS PAGE (with its blank side facing you).

Then, TURN THE PAGE.

The answers to the questions below are under the mask, opposite.

PROBLEM 1

When $t = 0$, $\epsilon^{at} =$ _____

(We will use ϵ^x rather than e^x for the exponential function.)

SLIDE THE MASK DOWN LEVEL WITH THE LINE AND CHECK YOUR ANSWER

PROBLEM 2

$\epsilon^{at} \rightarrow 0$ as $t \rightarrow \infty$ *provided* that _____

SLIDE THE MASK DOWN AGAIN TO CHECK YOUR ANSWER

PROBLEM 3

FIND the roots of the equation $s^2 + 2s + 2 = 0$

Answer: $s_1 =$ _____

$s_2 =$ _____

PROBLEM 4

EXPRESS $s^2 + 2s + 2$ as the product of two terms:

Answer: (_____)(_____) .

$\epsilon^{a0} = \underline{1}$

LEAVE THE MASK WHERE IT IS, AND ANSWER PROBLEM 2

$\epsilon^{at} \to 0$ as $t \to \infty$ provided that $a < 0$ or $\underline{a}$ is negative.

Now, GO TO **PROBLEM 3**

Using the quadratic formula,

$$s_1, s_2 = \frac{1}{2}\{-2 \pm \sqrt{2^2 - 4(2)}\}$$

$$s_1 = \underline{-1 + j1} \qquad \text{(We will use } j \text{ instead of } i \text{ for } \sqrt{-1})$$

$$\text{and} \quad s_2 = \underline{-1 - j1}$$

$$s^2 + 2s + 2 = (s - s_1)(s - s_2)$$

$$= \underline{(s + 1 - j1)(s + 1 + j1)} \longleftarrow \text{watch the signs!}$$

AGAIN, INSERT THE MASK BEHIND THIS PAGE.
THEN, TURN THE PAGE.

PROBLEM 5

SOLVE $\quad\left.\begin{array}{r} 2x + 3y = 5 \\ x - 2y = -8 \end{array}\right\}$ for the unknown y,

using determinants (Cramer's rule):

Answer: $y =$ _____

PROBLEM 6

COMPLETE the long division

$$s + 2 \overline{\smash{\big)}\, \begin{array}{c} s^2 \\ \hline s^3 + 2.5s^2 + 0.0s - 2 \end{array}}$$

PROBLEM 7

$\dfrac{d}{dt}(\epsilon^{at}) =$ _____ $\qquad$ and $\qquad$ $\dfrac{d}{dt}(\cos \omega t) =$ _____

PROBLEM 8

$\displaystyle\int x\,dx =$ _____ $\qquad$ and $\qquad$ $\displaystyle\int_0^t x\,dx =$ _____

PROBLEM 9

$\displaystyle\int \epsilon^{at}\,dt =$ _____ $\qquad$ and $\qquad$ $\displaystyle\int \cos \omega t\,dt =$ _____

$$y = \frac{\begin{vmatrix} 2 & 5 \\ 1 & -8 \end{vmatrix}}{\begin{vmatrix} 2 & 3 \\ 1 & -2 \end{vmatrix}} = \frac{-16 - 5}{-4 - 3} = \frac{-21}{-7} = \underline{3}$$

$$
\begin{array}{r}
s^2 + 0.5s \;\; - \;\; 1 \\
s + 2\,\overline{\smash{\big)}\,s^3 + 2.5s^2 + 0.0s - 2} \\
\underline{s^3 + 2.0s^2 } \\
0.5s^2 + 0s \\
\underline{0.5s^2 + \;\; s } \\
- \;\; s \;\; - 2 \\
\underline{- \;\; s \;\; - 2} \\
0 \qquad 0
\end{array}
$$

$\underline{a\epsilon^{at}}$ and $\underline{-\omega \sin \omega t}$ (If we assume that a and ω are given *constants*.)

$\underline{x^2/2 + C}$ and $\underline{x^2/2 \big|_0^t \;\; or \;\; t^2/2}$

$\underline{\dfrac{1}{a}\,\epsilon^{at} + C}$ and $\underline{\dfrac{1}{\omega}\,\sin \omega t + C}$

MOVE THE MASK AS BEFORE AND THEN TURN THE PAGE.

PROBLEM 10

Given that $\int u\,dv = uv - \int v\,du,$

EVALUATE $\int t\epsilon^t dt$ by integrating by parts:

Answer: _____

PROBLEM 11

MARK the point corresponding
to the complex number

$$-3 + j3$$

on the adjacent diagram.
(We will use j to represent $\sqrt{-1}$.)

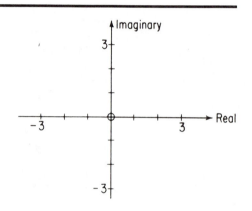

PROBLEM 12

If $-3 + j3$ is expressed in the polar (exponential) form, $M\epsilon^{j\theta}$,

$M = $ _____

and $\theta = $ _____ degrees

PROBLEM 13

The complex conjugate of $-3 + j3$ is _____

Let $u = t$ and $dv = \epsilon^t dt$

then $du = dt$ and $v = \int \epsilon^t dt = \epsilon^t$

thus $uv = t\epsilon^t$ and $\int v\,du = \int \epsilon^t dt = \epsilon^t$

and $\int t\epsilon^t dt = \underline{t\epsilon^t - \epsilon^t + C}$

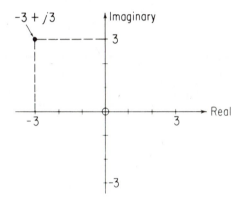

From the graph,

$$M = \underline{\sqrt{3^2 + 3^2}} \quad \text{or} \quad \underline{3\sqrt{2}}$$

and $\quad \theta = 90° + \tan^{-1}\dfrac{3}{3} = \underline{135°}$

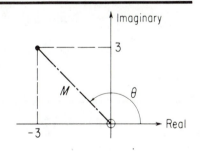

$\underline{-3 - j3}$ (simply replace j with $-j$)

KEEP MOVING THE MASK AHEAD AS YOU FINISH EACH PAGE.

PROBLEM 14

RATIONALIZE the ratio $\dfrac{1}{3-j}$ $\left(\text{that is, express } \dfrac{1}{3-j} \text{ in the form } A + jB\right).$

PROBLEM 15

MARK the point corresponding
to the complex number

$$-6 - j8$$

on the adjacent diagram.

Now, EXPRESS this number
in polar (exponential) form.

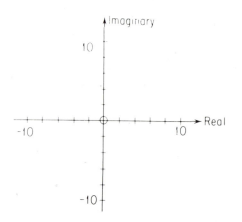

PROBLEM 16

Given that $\epsilon^{j\theta} = \cos\theta + j\sin\theta$,

$$\epsilon^{-j\theta} = \underline{\hspace{4cm}}$$

PROBLEM 17

Using the results of **PROBLEM 16**
EXPRESS $\cos\theta$ in terms of $\epsilon^{j\theta}$ and $\epsilon^{-j\theta}$:

$$\frac{1}{3-j} = \frac{1}{3-j}\frac{3+j}{3+j} = \frac{3+j}{9+1} = \underline{\frac{3}{10} + j\frac{1}{10}}$$

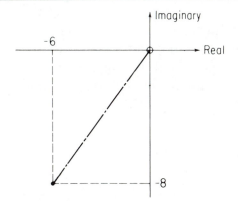

$$-6 - j8 = \sqrt{6^2 + 8^2}\,\epsilon^{-j(90^\circ + \tan^{-1}6/8)}$$

$$= \underline{10\epsilon^{-j126^\circ.8}} \quad \text{or equivalent, such as } 10\epsilon^{j233^\circ.2}$$

$$\epsilon^{-j\theta} = \underline{\cos\theta - j\sin\theta}$$

$$\epsilon^{j\theta} + \epsilon^{-j\theta} = 2\cos\theta$$

$$\cos\theta = \underline{\frac{1}{2}\{\epsilon^{j\theta} + \epsilon^{-j\theta}\}}$$

You have reached the end of the test of prerequisites. If you have had trouble solving *any* of the problems, *stop*, and see your instructor and/or review the topic which caused your difficulty.

The rest of the program follows a similar pattern—work on the left, then check your results on the right. If many of the frames seem easy, don't use this as an excuse to skip ahead; the sequencing is critical. If, on the other hand, the frames seem difficult and you find yourself making a lot of errors, *stop*, and see your instructor. (Not counting errors caused by carelessness, you should not make more than one error every ten frames or so.)

If you work through the program in the prescribed manner you *will* learn how to solve ordinary linear differential equations by the Laplace transform method. Trust the program; it will give you adequate practice in all the important techniques, and it will provide you with periodic review. In addition, you will have the opportunity to test yourself by completing a sample exam at the end.

Program working time: Seven to fifteen hours, plus two hours for the sample test.

GROUND RULES

1. Work sequentially, do *not* skip or browse.
   (You may refer *back* to earlier frames at any time.)
2. *Always write* down your answer *before* you look under the mask.
   (*If* you are *completely* stuck, don't hestiate to look at the answer for help.)
3. *If your answer is wrong*, partially wrong, or incomplete, or if you needed help,

       (a) Mark the frame with an "X" in the box provided; and
       (b) Correct or complete your answer with a red pencil.

   (This will help you track down difficulties when you review your work.)

SLIP THE MASK UNDER PAGE 1, OPPOSITE, AND CONTINUE TO READ.

When using logarithms to multiply two numbers we convert the process of multiplication into one of addition. The procedural steps are:
1. *Transform* the numbers into their logarithms;
2. Perform the operation of addition on the logarithms; and
3. Take the antilog of the result of step 2 to obtain the answer.

When we use the Laplace Transformation, we convert the process of solving differential equations (D.E.'s) into that of solving algebraic equations. In this case the steps are:
1. *Transform* the D.E.'s into algebraic equations;
2. Solve these equations for the algebraic unknowns; and
3. Take the inverse transform of the results of step 2 to obtain the answer.

It is this process that the program will develop. After a preliminary mathematical skirmish we will find that the transformation can be accomplished by "table look-up," just as you now look up the logarithm of a number. Then, after solving the *algebraic* equation, the solution of the D.E. follows from another table look-up, just as you might look up an antilogarithm.

TURN THE PAGE to start the first section provided that you are free to work for 30 to 50 minutes without interruption.

If you did not start with the section entitled **TO THE STUDENT**, or *if* you have not worked the test of prerequisites, GO BACK and do so.

1.1   As a matter of notation, we usually write log(a) when we "take the logarithm of a."

Analogously, we use $\mathcal{L}[\ \ ]$ to represent the process of "taking the Laplace Transform."

Thus the Laplace Transform (L.T.) of a function, $f(t)$, is written _____

AFTER YOU HAVE WRITTEN IN YOUR ANSWER,
SLIDE THE MASK DOWN TO THE LINE.
IF YOU ARE IN ERROR, PUT AN "X" IN THE BOX. $\longrightarrow$

1.2   In a more specific case we would write the L.T. of cos ωt as _____

1.3   We know how to evaluate log(a) but we do not as yet know how to evaluate $\mathcal{L}[f(t)]$.

To this end we *define*

$$\mathcal{L}[f(t)] = \int_0^\infty f(t)\,\epsilon^{-st}\,dt,$$

where ϵ^{-st} is the notation for the exponential function, exp($-st$), and s is a new (algebraic) variable.

As justification we will go on to show that this process has the desired property of transforming D.E.'s in time, t, into algebraic equations in the new variable, s.

Now, TURN THE MASK OVER and COPY the above definition *carefully* onto line 1 of the semi-notes. These semi-notes will serve as a record of important results. You may refer to them freely.

SLIP THE MASK UNDER PAGE 3
AND READ THE COMMENT THEREON $\rightarrow$

2

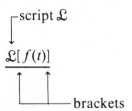

script $\mathcal{L}$

$\underline{\mathcal{L}[\,f(t)\,]}$

brackets

$\underline{\mathcal{L}[\cos\omega t]}$

Comment:

Don't let the definition scare you. Think of it in terms of the mechanical processes of

    i) multiplying $f(t)$ by the exponential ϵ^{-st},

  ii) integrating the product with respect to t, and

 iii) substituting the limits 0 and ∞

TURN THE PAGE

1.4 *Now* we can write

$$\mathcal{L}[\cos \omega t] = \int_0^\infty \underline{\hspace{4cm}} \epsilon^{-st} \, dt$$

(If necessary, refer to your semi-notes on the mask, they are there to help you.)

1.5 Similarly, $\mathcal{L}[t^n] = \int_0^\infty t^n \underline{\hspace{2cm}} dt$

1.6 To evaluate the L.T. of ϵ^{at} we would write:

$$\mathcal{L}[\epsilon^{at}] = \int_0^\infty \underline{\hspace{5cm}} dt$$

1.7 And, in the same vein:

$$\mathcal{L}\left[\frac{df(t)}{dt}\right] = \underline{\hspace{6cm}}$$

$$\int_0^\infty \underline{\cos \omega t \ \epsilon^{-st}} \, dt$$

$$\int_0^\infty t^n \underset{\underline{\quad}}{\overset{\overset{\text{—note sign}}{\downarrow}}{\epsilon^{-st}}} \, dt$$

$$\int_0^\infty \underline{\epsilon^{at} \, \epsilon^{-st}} \, dt \qquad \text{or} \qquad \int_0^\infty \underline{\epsilon^{(a-s)t}} \, dt$$

Comment:

Alternative algebraic equivalents may always be accepted as correct—as long as you are sure that they *are* equivalent.

$$\underline{\int_0^\infty \frac{df(t)}{dt} \, \epsilon^{-st} \, dt} \qquad \text{or} \qquad \underline{\int_0^\infty \epsilon^{-st} \, df(t)}$$

Comment:

$\dfrac{df(t)}{dt}$ is a perfectly respectable function of time!

1 8   The Laplace Transformation,

$$\mathcal{L}[f(t)] = \int_0^\infty f(t)\,\epsilon^{-st}\,dt$$

introduces a new variable, s, which we *define* to be independent of t.

WRITE "where s is a new variable, which is <u>independent</u> of t" on line 2 of your semi-notes.

1.9   If we are to evaluate, for example:

$$\mathcal{L}[\cos \omega t] = \int_0^\infty \cos \omega t\, e^{-st}\,dt$$

the quantity s should be regarded as _____ while integrating with respect to t.

□

> **IMPORTANT NOTE**
>
> A program is not an exam, even though you are answering questions. You have nothing to gain—and you may fail to learn—if you look at the answers before you have *written* down your response.

a constant or fixed or constant

(since s is independent of the variable of integration, t.)

1.10   Now let's *find* the L.T. of $A\epsilon^{at}$, where A and a are given *constants*. From the definition,

$$\mathcal{L}[A\epsilon^{at}] = \int \underline{\hspace{6cm}}$$

(Don't try to evaluate the integral in this frame.)

1.11   When we integrate we will have to evaluate $\epsilon^{(a-s)t}$ at the upper limit of $t \rightarrow \infty$. If we assume that a and s are real, and that s may be chosen so that $(a - s) < 0$, it follows that

$$\epsilon^{(a-s)\infty} = \underline{\hspace{4cm}}$$

1.12   Now you should be able to EVALUATE

$$\mathcal{L}[A\epsilon^{at}] = A \int_0^\infty \epsilon^{(a-s)t}\, dt \qquad (A \text{ and } a \text{ are } constants)$$

$$= $$

$$= $$

$$= \underline{\hspace{2cm}}/(\underline{\hspace{2cm}})$$

READ THE COMMENTS OPPOSITE

$$\mathcal{L}[A\epsilon^{at}] = \int_0^\infty A\epsilon^{at}\epsilon^{-st}\,dt \qquad \text{or} \qquad A\int_0^\infty \epsilon^{(a-s)t}\,dt$$

_____

(or equivalent)

$\epsilon^{(a-s)\infty} = \underline{0}$ since $(a - s) < 0$.

NOTE:

The *format* of the *following* frame is intended to indicate:
1. That *several* mathematical steps will be required (not necessarily the *same* number of steps as there are equal signs); and
2. That the final step should leave the result in some specified form (here the ratio of two quantities).

$$\mathcal{L}[A\epsilon^{at}] = A\int_0^\infty \epsilon^{(a-s)t}\,dt$$

$$= \left.\frac{A\epsilon^{(a-s)t}}{(a-s)}\right|_0^\infty$$

$$= \frac{A}{(a-s)}\{\epsilon^{(a-s)\infty} - \epsilon^{(a-s)0}\}$$

$$= \frac{A}{(a-s)}\{0 - 1\}$$

$$\boxed{\mathcal{L}[A\epsilon^{at}] = A/(s-a)} \qquad \text{or} \qquad -A/(a-s)$$

(a is a given constant, and s is independent of t. Therefore $(a - s)$ is regarded as a constant in the integration with respect to t.)

(the first term in the braces, $\epsilon^{(a-s)\infty}$, is zero *provided* that $(a - s)$ is less than zero, as assumed.)

Comments:
1. a and s were assumed to be real in order to simplify the analysis. The result is valid for any a, real or complex, provided that s is chosen so that $\Re e\{a - s\} < 0$.
2. At this time we will simply assume that s can be so chosen.

1.13   Just as we must prepare a table of logarithms before we can make convenient numerical use of the log function, a table of "transform pairs" is an important adjunct to the application of the L.T. method. We will build up such a table on our semi-notes.

We derived the result, $\mathcal{L}[A\epsilon^{at}] = A/(s - a)$, on the previous page. You may now COPY the first "transform pair,"

$$A\epsilon^{at} \leftrightarrow A/(s - a)$$

onto line 5 of your semi-notes. Then READ the comment opposite.

1.14   As an example of the application of a table of transform pairs, suppose we have a voltage source whose terminal voltage, $v(t)$, varies according to the relation $v(t) = 7\epsilon^{-2t}$:

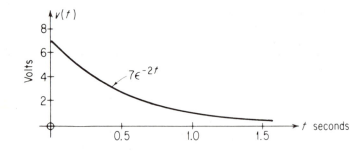

You can determine, from line 5 of your semi-notes, that:

$$\mathcal{L}[v(t)] = \underline{\hspace{3cm}}$$

1.15   A second transform pair follows from the first if we put $a = 0$ in the result

$$\mathcal{L}[A\epsilon^{at}] = A/(s - a).$$

That is, putting $a = 0$ yields the special case

$$\mathcal{L}[\underline{\hspace{2cm}}] = \underline{\hspace{2cm}}$$

Comment:
$A\epsilon^{at}$ should be in the left-hand column of the table, under $f(t)$; $A/(s-a)$ should be in the right-hand column, under the heading $\mathcal{L}[f(t)]$

$$\mathcal{L}[v(t)] = \mathcal{L}[7\epsilon^{-2t}] = \underline{7/(s+2)}$$

└─note sign

Comment:
We could, of course, obtain the same result by integration:

$$\mathcal{L}[v(t)] = \mathcal{L}[7\epsilon^{-2t}] = 7\int_0^\infty \epsilon^{-2t}\epsilon^{-st}\,dt = \cdots$$

The table of transform pairs is thus no more than a timesaving device.

If $\quad\quad\mathcal{L}[A\epsilon^{at}] = A/(s-a)$

Then $\quad\quad\mathcal{L}[A\epsilon^{0t}] = A/(s-0) \quad\quad$ (putting $a = 0$)

or $\quad\quad\boxed{\mathcal{L}[A] = A/s}$

1.16   COPY the new transform pair,

$$A \leftrightarrow A/s$$

onto line 6 of your semi-notes.

1.17   When you derived the result $\mathcal{L}[A\epsilon^{at}] = A/(s-a)$

or $\mathcal{L}[7\epsilon^{-2t}] = 7/(s+2)$

or $\mathcal{L}[A] = A/s$

You transformed a function of t into a function of _____ .

1.18   It has become commonplace to employ the notation

$$\mathcal{L}[f(t)] = F(s).$$

In the same notation,

$$\mathcal{L}[g(t)] = \underline{\hspace{2cm}}$$

1.19   ADD to line 1 of your semi-notes as follows:

$$\mathcal{L}[f(t)] = \int_0^\infty f(t)\,\epsilon^{-st}\,dt = F(s)$$

└──addition

1.20   Similarly, we may represent the L.T. of a current, $i(t)$, by

$$\mathcal{L}[i(t)] = \underline{\hspace{2cm}}$$

and the L.T. of a velocity, $v(t)$, by _____ .

The L.T. of a function of t is a *function of s*; a "change of variable."

(t "disappears" since it is everywhere replaced by the limits of integration, 0 and ∞.)

capital

$$\mathcal{L}[g(t)] = \underline{\dot{G}(s)} \leftarrow \text{function of } s$$

Comment:
$F(s)$ is no more than a compact and therefore convenient way of writing $\mathcal{L}[f(t)]$. The two forms are mathematically equivalent, although the notation $F(s)$ emphasizes the fact that the transform is a *function of s*.

$$\mathcal{L}[i(t)] = \underline{I(s)}$$

and the L.T. of $v(t)$ is

$$\mathcal{L}[v(t)] = \underline{V(s)}$$

1.21   As review, use your table of transform pairs to

EVALUATE $\mathcal{L}[10] =$ _____ .

$\square$

1.22   Similarly, $\mathcal{L}[100\epsilon^{2t}] =$ _____ .

$\square$

1.23   And, $\mathcal{L}[10 + 100\epsilon^{2t}] =$ _____ .

(Try this, even if you feel you are "guessing" a little.)

$\square$

1.24   To justify the last result, consider the following, where a and b are *constants*, and $f(t)$ and $g(t)$ are two functions of time:

$$\mathcal{L}[af(t) + bg(t)] = \int_0^\infty \{af(t) + bg(t)\}\epsilon^{-st}\,dt$$

$$= \int_0^\infty \{af(t)\epsilon^{-st} + bg(t)\epsilon^{-st}\}\,dt$$

$$= a\int_0^\infty f(t)\epsilon^{-st}\,dt + b\int_0^\infty g(t)\epsilon^{-st}\,dt$$

$$\boxed{\mathcal{L}[af(t) + bg(t)] = aF(s) + bG(s)}$$

COMPLETE line 12 of your semi-notes (Theorem 1).

14

$$\mathcal{L}[10] = \underline{10/s} \qquad \text{(line 6 of semi-notes)}$$

Use your semi-notes freely for reference.

$$\mathcal{L}[100\epsilon^{2t}] = \underline{100/(s-2)} \qquad \text{(line 5 of semi-notes)}$$

$$\mathcal{L}[10 + 100\epsilon^{2t}] = \mathcal{L}[10] + \mathcal{L}[100\epsilon^{2t}] = \underline{10/s + 100/(s-2)}$$

(This assumes, correctly, that *the L.T. of a sum is equal to the sum of the transforms of the separate terms.*)

1.25   Given $x(t) = 5 - 4\epsilon^{-2t}$,

$$X(s) = \underline{\hspace{5cm}}.$$

☐

1.26   As a special case of Theorem 1, we may conclude that:

$$\mathcal{L}[Cz(t)] = \underline{\hspace{4cm}}$$

where C is a constant.

☐

1.27   In a similar way,

$$\mathcal{L}[v(t)/R] = \underline{\hspace{4cm}}.$$

☐

REVIEW

1.28   *Without* looking at your semi-notes or earlier frames, COMPLETE the *definition* of the L.T.:

$$\mathcal{L}[f(t)] = \int \underline{\hspace{5cm}}$$

☐

1.29   The L.T. of a function of t is a function of $\underline{\hspace{2cm}}$.

☐

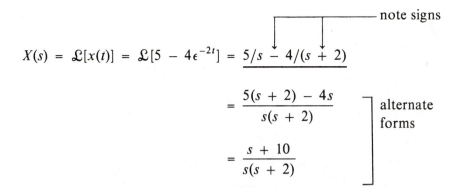

$$X(s) = \mathcal{L}[x(t)] = \mathcal{L}[5 - 4\epsilon^{-2t}] = \underline{5/s - 4/(s + 2)}$$

$$= \frac{5(s + 2) - 4s}{s(s + 2)} \quad \left.\begin{array}{l} \text{alternate} \\ \text{forms} \end{array}\right.$$

$$= \frac{s + 10}{s(s + 2)}$$

Putting $b = 0$ in Theorem 1 we have the special case:

$$\mathcal{L}[af(t)] = aF(s)$$

That is, the constant in $\mathcal{L}[Cz(t)]$ comes outside, giving:

$$\mathcal{L}[Cz(t)] = \underline{CZ(s)} \quad \text{or} \quad \underline{C\mathcal{L}[z(t)]}$$

$$\mathcal{L}[v(t)/R] = \mathcal{L}\left[\frac{1}{R} v(t)\right] = \underline{\frac{1}{R} V(s)}$$

Again, the constant $1/R$ "comes outside."

$$\mathcal{L}[f(t)] = \underline{\int_0^\infty f(t)\epsilon^{-st} \, dt}$$

a function of $\underline{s}$.

SUMMARY

So far we have
1. *Defined* the L.T. of $f(t)$.
2. Developed the notations $\mathcal{L}[f(t)] = F(s)$ for the L.T. of $f(t)$.
3. Calculated $\mathcal{L}[A\epsilon^{at}]$ and $\mathcal{L}[A]$.
4. Established that: The L.T. of a sum equals the sum of the L.T.'s (Theorem 1).

When we defined the L.T. in frame **1.3** we stated that it had the property of transforming D.E.'s into algebraic equations.

Obviously we cannot demonstrate this characteristic until we learn how to take the L.T. of a derivative, which we will do in the next section.

Do not continue unless or until you have 15 to 20 minutes free of interruption.

LAPLACE TRANSFORM OF THE FIRST DERIVATIVE

2.1   We first return to the *definition* of the L.T. and write:

$$\mathcal{L}\left[\frac{df(t)}{dt}\right] = \underline{\hspace{8cm}}$$

2.2   To evaluate the integral above we integrate by parts, using the usual formula

$$\int_a^b u \, dv = uv \Big|_a^b - \int_a^b v \, du.$$

Thus

$$\mathcal{L}\left[\frac{df(t)}{dt}\right] = \int_0^\infty \epsilon^{-st} \frac{df(t)}{dt} \, dt$$

$$= \epsilon^{-st} f(t)\Big|_0^\infty - \int_0^\infty f(t)\{-s\epsilon^{-st} \, dt\}$$

Comparing these two equations:

$u = \underline{\hspace{3cm}}$; $v = \underline{\hspace{3cm}}$; $a = \underline{\hspace{3cm}}$;

$du = \underline{\hspace{3cm}}$; $dv = \underline{\hspace{3cm}}$; $b = \underline{\hspace{3cm}}$.

2.3   Proceeding,

$$\mathcal{L}\left[\frac{df(t)}{dt}\right] = \int_0^\infty \epsilon^{-st} \frac{df(t)}{dt} \, dt = \epsilon^{-st} f(t)\Big|_0^\infty - \int_0^\infty f(t)\{-s\epsilon^{-st} \, dt\}$$

$$= \epsilon^{-s\infty} f(\infty) - \epsilon^{-s0} f(0) + s \int_0^\infty f(t)\epsilon^{-st} \, dt$$

$$= \epsilon^{-s\infty} f(\infty) - f(0) + sF(s)$$

For that class of functions for which $\epsilon^{-st} f(t) \to 0$ as $t \to \infty$ (for some finite s) it follows that,

$$\boxed{\mathcal{L}\left[\frac{df(t)}{dt}\right] = sF(s) - f(0)}$$

COPY this result onto line 13 of your semi-notes as Theorem 2, and then *READ the comment on page 21, opposite.*

20

$$\mathcal{L}\left[\frac{df(t)}{dt}\right] = \int_0^\infty \frac{df(t)}{dt}\, \epsilon^{-st}\, dt \qquad \text{or} \qquad \int_0^\infty \epsilon^{-st}\, \frac{df(t)}{dt}\, dt$$

(alternate forms)

Comment:
We took this step earlier, in frame **1.7**

$$\int_0^\infty \epsilon^{-st}\, \frac{df(t)}{dt}\, dt = \epsilon^{-st} f(t)\Big|_0^\infty - \int_0^\infty f(t)\{-s\epsilon^{-st}\, dt\}$$

$$\int_a^b u\, dv \qquad = \qquad uv\Big|_a^b - \int_a^b v\, du$$

$$u = \epsilon^{-st}; \qquad v = f(t); \qquad\qquad a = 0;$$

$$du = -s\epsilon^{-st}\, dt; \quad dv = \frac{df(t)}{dt}\, dt \;\; \text{or} \;\; df(t); \quad b = \infty.$$

note

Comment:
All the "common engineering functions," such as ϵ^{at}, $\cos \omega t$, $\cosh at$, t^n, and products of these, fall into the class for which Theorem 2 is applicable.

Even though some functions in this class become infinite as $t \to \infty$, we can always choose a finite s such that $\epsilon^{-st} \to 0$ more rapidly than $f(t) \to \infty$. The *product* $\epsilon^{-s\infty} f(\infty)$ therefore tends to zero.

There are, however, some functions, such as ϵ^{t^2}, for which $\epsilon^{-s\infty} f(\infty) \to \infty$ for all finite values of s. $\mathcal{L}[df(t)/dt]$ does not exist for these functions, and Theorem 2 may not then be used.

2.4   If $\mathcal{L}\left[\dfrac{df(t)}{dt}\right] = sF(s) - f(0)$

we would write

$$\mathcal{L}\left[\dfrac{dv(t)}{dt}\right] = \underline{\hspace{5cm}}$$

2.5   Then, using Theorems 1 and 2 together,

$$\mathcal{L}\left[m\,\dfrac{dv(t)}{dt}\right] = \underline{\hspace{4cm}}$$

where m is a constant.

2.6   Similarly, $\mathcal{L}\left[L\,\dfrac{di(t)}{dt}\right] = \underline{\hspace{4cm}}$

where L is a constant.

2.7

Initial condition:
Initial value of $i(t)$ is $+5$ amp.

The voltage-current relation for an inductor is:

$$v(t) = L\,\dfrac{di(t)}{dt}$$

Transforming, $V(s) = \underline{\hspace{4cm}} = \underline{\hspace{3cm}}$ (substituting values)

$$\mathcal{L}\left[\frac{dv(t)}{dt}\right] = \underline{sV(s) - v(0)}$$

$$\mathcal{L}\left[m\,\frac{dv(t)}{dt}\right] = m\,\mathcal{L}\left[\frac{dv(t)}{dt}\right] = \underline{m\{sV(s) - v(0)\}} \qquad \text{or} \qquad \underline{msV(s) - mv(0)}$$

(the constant m comes outside, from Theorem 1.)

$$\mathcal{L}\left[L\,\frac{di(t)}{dt}\right] = \underline{L\{sI(s) - i(0)\}} \qquad \text{or} \qquad \underline{LsI(s) - Li(0)}$$

If $\qquad v(t) = L\,\dfrac{di(t)}{dt}$

Then $\quad V(s) = L\{sI(s) - i(0)\} = \underline{2\{sI(s) - 5\}} \qquad$ or $\qquad \underline{2sI(s) - 10}$

Comment:
$i(0)$ is the value of $i(t)$ at $t = 0$, such value being called the "initial value of $i(t)$" or the "*initial condition.*"

2.8   Finally, TRANSFORM the differential equation:

$$y(t) = 3x(t) + 7\frac{dx(t)}{dt} \qquad \text{with initial condition } x(0) = 4.$$

$$Y(s) = \underline{\hspace{8cm}}$$

$$\boxed{}$$

SUMMARY

We are now able to transform a first derivative into an algebraic expression. For example:

$$\mathcal{L}\left[\frac{d}{dt}f(t)\right] = sF(s) - f(0)$$

The operation of differentiation is transformed into the algebraic operation of multiplication by s, together with the subtraction of the constant, $f(0)$, which is the initial value of $f(t)$, that is, the value of $f(t)$ at $t = 0$.

In the next section we will solve a first order differential equation.
Continue only when you have 15 to 40 minutes available.

$$Y(s) = \mathcal{L}\left[3x(t) + 7\frac{dx(t)}{dt}\right]$$

$$= \underline{3X(s) + 7\{sX(s) - x(0)\}}$$

$$\left.\begin{array}{l} = 3X(s) + 7\{sX(s) - 4\} \\ \\ = 3X(s) + 7sX(s) - 28 \end{array}\right] \quad \text{alternatives}$$

SOLUTION OF CIRCUIT PROBLEM I

3.1   With the necessary preliminaries behind us, let's try our hand at solving a circuit problem:

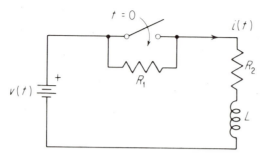

Switch open for $t < 0$
Switch *closed* for $t \geq 0$

Our aim is to find $i(t)$ for $t \geq 0$; that is, after the switch has closed. It is easy to see from Kirchhoff's voltage law that for $t \geq 0$,

$$L \frac{di(t)}{dt} + R_2 i(t) = v(t) \qquad \text{with an initial condition, } i(0).$$

3.2   We agreed to represent the L.T. of $f(t)$ by $F(s)$. Similarly,

$$\mathcal{L}[i(t)] = \underline{\hspace{3cm}}$$

and $\mathcal{L}[v(t)] = \underline{\hspace{3cm}}$

3.3   Also, we can write

$$\mathcal{L}[R_2 i(t)] = \underline{\hspace{3cm}}$$

and $\mathcal{L}\left[L \frac{di(t)}{dt}\right] = \underline{\hspace{5cm}}$ (You may use your semi-notes)

3.4   We can now TRANSFORM the D.E. in frame **3.1** above, term by term, to yield the *transformed* equation:

Comment:
The *formulation* of the D.E.'s is *not* a part of this program.   Physical systems, such as the electric circuit opposite, are offered to breathe some "life" into the equations.

If for some reason you do not understand the formulation (either here or in later examples) you should simply accept the differential equation and proceed with its *solution*.

$$\mathcal{L}[i(t)] = \underline{I(s)}$$

and

$$\mathcal{L}[v(t)] = \underline{V(s)}$$

$$\mathcal{L}[R_2 i(t)] = \underline{R_2 I(s)}$$

$$\mathcal{L}\left[L \, \frac{di(t)}{dt}\right] = \underline{L\{sI(s) - i(0)\}} \qquad \text{(or equivalent)}$$

— note

Given $L \, \dfrac{di(t)}{dt} + R_2 i(t) = v(t)$, with initial condition, $i(0)$,

then

$$\underline{L\{sI(s) - i(0)\}} + \underline{R_2 I(s)} = \underline{V(s)}$$

27

3.5   Suppose that we are given

$$v(t) = \text{constant} = K$$

Then, in terms of K,

$$V(s) = \underline{\hspace{3cm}}$$

3.6   Suppose that we are also given*

$$R_2 = 2 \text{ ohm}, \qquad L = 1 \text{ henry}, \qquad i(0) = 1 \text{ ampere},$$

and $K = 10 \text{ volt}, \quad \text{or} \quad V(s) = 10/s$

If we SUBSTITUTE the above data in the transformed equation

$$L\{sI(s) - i(0)\} + R_2 I(s) = V(s)$$

we obtain

$$\underline{\hspace{8cm}} = \underline{\hspace{2cm}}$$

---

*The numerical values used in this text have been chosen to simplify your calculations and are not necessarily "practical" values.

If $v(t) = K$, a constant

then $V(s) = \mathcal{L}[v(t)] = \underline{K/s}$ (line 6 of your semi-notes)

By simple substitution,

$$1\{sI(s) - 1\} + 2I(s) = 10/s \quad \text{or} \quad sI(s) - 1 + 2I(s) = 10/s$$

3.7   We have transformed a D.E. in the unknown, $i(t)$, into an algebraic equation with $I(s)$ the unknown. As the next step,

SOLVE the equation

$$sI(s) - 1 + 2I(s) = 10/s$$

for $I(s)$:

3.8   With $I(s)$ now known, we have only to "take the inverse transform" to find $i(t)$.

It is easily verified that we can write

$$I(s) = \frac{s + 10}{s(s + 2)} \equiv \frac{5}{s} - \frac{4}{s + 2}$$

and your table of transform pairs should then suggest that:

$i(t) = $ _____ amp

3.9   Before we congratulate ourselves, let's make a quick check. SUBSTITUTE $t = 0$ in the above expression for $i(t)$, yielding

$i(0) = $ _____ amp

Then CHECK this value against the initial condition (I.C.) given on frame **3.6**.

From $\quad sI(s) - 1 + 2I(s) = 10/s$

we have $\quad\quad (s + 2)I(s) = \dfrac{10}{s} + 1 = \dfrac{10 + s}{s} \quad\quad$ (rearranging)

Therefore,

$$I(s) = \frac{s + 10}{s(s + 2)} \quad\quad\quad \text{(dividing by } s + 2)$$

Alternate forms are acceptable, but this is the *preferred* form—no negative powers of s in numerator or denominator, and the highest powers of s with unity coefficients.

Examples: $\dfrac{5(s + 16)}{s(s + 3)}$; $\quad \dfrac{4(s^2 + 7s - 5)}{s^2(s - 2)}$

Comment:
Note that $I(s)$ *is* a function of s, namely $(s + 10)/s(s + 2)$

If $\quad I(s) = 5/s - 4/(s + 2)$

then $\quad i(t) = \underline{5 - 4\epsilon^{-2t}}$ amp

COMPARE with frame **1.25**. Here we have "taken the inverse transform" by simply working that problem "backwards."

$\quad\quad\quad i(0) = \underline{1}$ amp $\quad\quad$ (substitution of $t = 0$ into $i(t)$.)

This agrees with the given I.C. on frame **3.6**

3.10   We have solved the circuit problem posed in frame **3.1**, but let's "look" at the solution, $i(t) = 5 - 4\epsilon^{-2t}$.

First you may note that

$$di(t)/dt = \underline{\hspace{5cm}} \text{ amp/sec}$$

3.11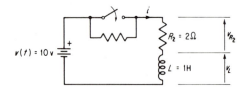

Now, using the information in frame **3.10** above and on the circuit diagram, COMPLETE the following table:

t second	$4\epsilon^{-2t}$	$i(t)$ amp	$v_{R_2} = R_2 i$ volt	$di(t)/dt$ amp/sec	$v_L = L di(t)/dt$ volt	$v_{R_2} + v_L$ volt
0.0						
0.5						
2.0						

(*Data:* $\epsilon^{-1} \doteq 0.37$ and $\epsilon^{-4} \doteq 0.02$)

3.12   Finally, SKETCH in the following graphs, thus illustrating the circuit's behavior after the closing of the switch.

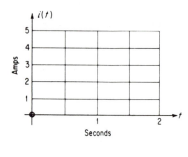

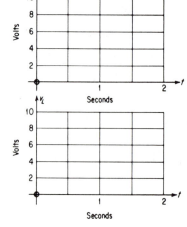

If $\qquad i(t) = 5 - 4\epsilon^{-2t}$ amp

then $\quad di(t)/dt = \underline{8\epsilon^{-2t}}$ amp/sec

---

t second	$4\epsilon^{-2t}$	$i(t)$ amp	$v_{R_2} = R_2 i$ volt	$di(t)/dt$ amp/sec	$v_L = L di(t)/dt$ volt	$v_{R_2} + v_L$ volt
0.0	4.00	1.00	2.00	8.00	8.00	10.0
0.5	1.48	3.52	7.04	2.96	2.96	10.0
2.0	0.08	4.92	9.84	0.16	0.16	10.0

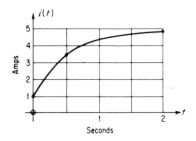

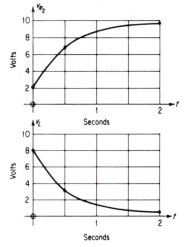

There is not much more to be known about *this* circuit!

SUMMARY

We started with a circuit (frame **3.1**) in which the current $i(t)$ was unknown. The "solution" may be either a mathematical result, $i(t) = 5 - 4\epsilon^{-2t}$, *or* a graphical result. In this instance we have found both.

Given the D.E. describing the physical circuit, we proceeded to *solve* the D.E. using the method of the L.T., and following the steps proposed earlier. That is, we
1. *Transformed* the D.E. into an algebraic equation;
2. Solved this equation for the algebraic unknown; and
3. Took the inverse transform of the algebraic solution to find the corresponding time function.

An estimate of the working time is given for each section of the program so that you can arrange to terminate each work period at the *end* of a section. Also, it is a good idea to avoid working for more than an hour or so at one sitting.

The solution of the following problem will take you between 15 and 25 minutes.

SLIP THE MASK UNDER PAGE 35 AND CONTINUE TO READ.

REVIEW OF THE GROUND RULES

1. Work sequentially, do *not* skip or browse.
   (You may refer *back* to earlier frames at any time.)
2. *Always write* down your answer *before* you look under the mask.
   (*If* you are *completely* "stuck," don't hesitate to look at the answer for help.)
3. *If your answer is wrong*, partially wrong, or incomplete, or if you needed help,
        a. Mark the frame with an "X" in the box provided; and
        b. Correct or complete your answer with a red pencil or the equivalent.
   (This will help you track down difficulties when you review your work.)

If you are starting a new work period here, re-read the SUMMARY on the last page.

SOLUTION OF CIRCUIT PROBLEM II

4.1   To consolidate our position let's try a variation on a theme—the switch now *opens* at $t = 0$:

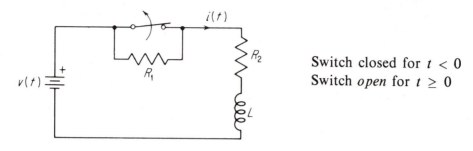

Switch closed for $t < 0$
Switch *open* for $t \geq 0$

The aim is again to find $i(t)$ for $t \geq 0$.

The circuit equation for $t \geq 0$ is only slightly different from the previous example:

$$L \frac{di(t)}{dt} + (R_1 + R_2)\, i(t) = v(t) \qquad \text{with an initial condition, } i(0).$$

This, when transformed term by term, yields:

If $\quad L \dfrac{di(t)}{dt} + (R_1 + R_2)\,i(t) = v(t)$

then $\quad L\{sI(s) - i(0)\} + (R_1 + R_2)I(s) = V(s)$

note the braces!

37

4.2   SUBSTITUTE the data

$$R_1 = 8\,\Omega, \quad R_2 = 2\,\Omega, \quad L = 1\text{H}, \quad i(0) = 5\,\text{amp}, \quad \text{and} \quad V(s) = 10/s$$

in the equation

$$L\{sI(s) - i(0)\} + (R_1 + R_2)I(s) = V(s)$$

and then SOLVE for $I(s)$:

Put your answer in the preferred form: $I(s) = \dfrac{(s + \quad)}{s(s + \quad)}.$

—a constant

Substituting the data:

$$sI(s) - 5 + 10I(s) = 10/s$$

rearranging,

$$(s + 10)I(s) = \frac{10}{s} + 5 = \frac{10 + 5s}{s}$$

whence

$$I(s) = \frac{10 + 5s}{s(s + 10)}$$

that is

$$I(s) = \frac{5(s + 2)}{s(s + 10)}$$

THE PARTIAL FRACTION EXPANSION

4.3   We would expect to be able to write

$$I(s) = \frac{5(s + 2)}{s(s + 10)} \equiv \frac{A}{s} + \frac{B}{s + 10}.$$

This process of decomposition, which is the reverse of combining terms over a common denominator, is called "expansion into Partial Fractions."

If we can find the values of the constants, A and B, the inverse transform, $i(t)$, will follow at once from our table of transform pairs.

How might we find A? One possibility would be to start by multiplying through by the denominator of the term in A (in this case, s):

$$\frac{5(s + 2)\cancel{s}}{\cancel{s}(s + 10)} \equiv A + \frac{Bs}{s + 10}$$

Then to complete the process we must eliminate the term in B, which can clearly be achieved (in this example) by putting $s = 0$:

$$A = \frac{5(s + 2)\cancel{s}}{\cancel{s}(s + 10)}\bigg|_{s=0} = \frac{10}{10} = 1 \qquad\qquad \text{SEE COMMENTS OPPOSITE}$$

4.4   Now FIND B, *using the same method.* That is, multiply through the original identity by B's denominator, $s + 10$, *then* set $s = -10$ to eliminate the term in A:

$$\boxed{}$$

4.5   VERIFY your expansion by recombining the two fractions over their common denominator:

$$\boxed{}$$

Comments:

1. $\dfrac{5(s + 2)}{s(s + 10)} \equiv \dfrac{A}{s} + \dfrac{B}{s + 10}$ is an *identity*, which must hold for *all* values of s. While computing A and B we are therefore at liberty to choose those particular values of s which simplify the calculation.

2. You may have previously learned a different method for calculating the constants in the Partial Fraction expansion. If so, *please* don't use it here. In some of the later problems you will get into difficulty if you do not use the procedure introduced on page 40, opposite.

Given $\quad \dfrac{5(s + 2)}{s(s + 10)} \equiv \dfrac{A}{s} + \dfrac{B}{(s + 10)} \qquad$ we find B as follows:

Step 1: Multiply through by B's denominator, $s + 10$

$$\frac{5(s + 2)\cancel{(s + 10)}}{s\cancel{(s + 10)}} = \frac{A(s + 10)}{s} + B$$

Step 2: Put $s = -10$ to eliminate the term containing A:

$$B = \left.\frac{5(s + 2)\cancel{(s + 10)}}{s\cancel{(s + 10)}}\right|_{s = -10} = \frac{-40}{-10} = \underline{4}$$

$$\frac{1}{s} + \frac{4}{s + 10} = \frac{s + 10 + 4s}{s(s + 10)} \qquad \text{(combining the terms over a common denominator)}$$

$$= \frac{5(s + 2)}{s(s + 10)} \qquad \text{(rearranging the numerator)}$$

which establishes the validity of the expansion.

4.6   Having found that

$$I(s) = 1/s + 4/(s + 10)$$

the solution follows as the inverse transform of $I(s)$:

$i(t) = $ 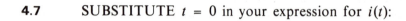 amp

---

4.7   SUBSTITUTE $t = 0$ in your expression for $i(t)$:

$i(0) = $ _____ amp

CHECK this result against the initial condition (I.C.) given in frame **4.2**.

This completes the solution of the circuit problem, but before we leave this section let's try another Partial Fraction expansion for practice in this procedure.

$$i(t) = \underline{1 + 4\epsilon^{-10t}} \text{ amp}$$

$$i(0) = \underline{5} \text{ amp} \qquad (\text{substituting } t = 0)$$

This agrees with the given $i(0)$ of frame **4.2**.

4.8   FIND A in the expansion

$$\frac{12s^2 + 22s + 6}{s(s + 1)(s + 2)} \equiv \frac{A}{s} + \frac{B}{s + 1} + \frac{C}{s + 2}.$$

(Use the method introduced earlier in this section.)

___

4.9   FIND B and C, using the same method:

___

The solution time for the next problem will be between 10 and 20 minutes.

To find A:
1. Multiply by the denominator of the term in A, and
2. Set s equal to the value which will cause the terms in B and C to vanish. Thus:

$$\frac{(12s^2 + 22s + 6)\cancel{s}}{\cancel{s}(s + 1)(s + 2)} = A + \frac{Bs}{s + 1} + \frac{Cs}{s + 2}$$

$$A = \left.\frac{(12s^2 + 22s + 6)\cancel{s}}{\cancel{s}(s + 1)(s + 2)}\right|_{s=0} = \frac{6}{2} = \underline{3}$$

By now you should be able to write down directly:

$$B = \left.\frac{(12s^2 + 22s + 6)\cancel{(s + 1)}}{s\cancel{(s + 1)}(s + 2)}\right|_{s=-1} = \frac{12 - 22 + 6}{-1(1)} = \frac{-4}{-1} = \underline{4}$$

and

$$C = \left.\frac{(12s^2 + 22s + 6)\cancel{(s + 2)}}{s(s + 1)\cancel{(s + 2)}}\right|_{s=-2} = \frac{48 - 44 + 6}{-2(-1)} = \frac{10}{2} = \underline{5}$$

Comment:
Please get into the habit of using this method *and* notation.

A LEAKY RESERVOIR

5.1 For additional practice we will solve for the water level, $h(t)$, in a cylindrical reservoir whose cross-sectional area is A square meters:

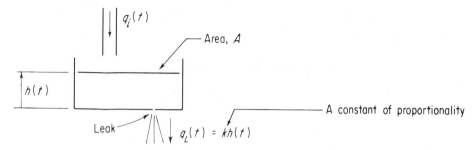

By conservation of mass,

$$A\,\frac{dh(t)}{dt} + kh(t) = q_i(t) \qquad \text{with an I.C. of } h(0)$$

Therefore, TRANSFORMING term by term, the *transformed* equation is:

_____   =   _____

☐

5.2. SUBSTITUTE the data

$$k = \tfrac{1}{2}\,m^2/\text{sec.}, \quad A = \tfrac{1}{4}m^2, \quad h(0) = 4m, \quad \text{and} \quad q_i(t) = 4\epsilon^{-t}\,m^3/\text{sec}$$

in the transformed equation above, and SOLVE for $H(s)$:

Put your answer in the preferred form $H(s) = \dfrac{(s + \quad)}{(s + \quad)(s + \quad)}$

If
$$\underline{A\,dh(t)/dt\; +\; kh(t)\; =\; q_i(t)}$$

then
$$\underline{A\{sH(s)\; -\; h(0)\}\; +\; kH(s)\; =\; Q_i(s)}$$

$$Q_i(s)\; =\; \mathcal{L}[4\epsilon^{-t}]\; =\; 4/(s\; +\; 1)\;\text{ and, by substitution,}$$

$$\tfrac{1}{4}\{sH(s)\; -\; 4\}\; +\; \tfrac{1}{2}H(s)\; =\; 4/(s\; +\; 1)$$

$$\tfrac{1}{4}(s\; +\; 2)H(s)\; =\; \frac{4}{s\; +\; 1}\; +\; 1\; =\; \frac{s\; +\; 5}{s\; +\; 1}$$

$$\therefore H(s)\; =\; \underline{\frac{4(s\; +\; 5)}{(s\; +\; 1)(s\; +\; 2)}}$$

47

5.3   If you wish, REVIEW the procedure for Partial Fraction expansion as summarized by the example of frames **4.8** and **4.9**. Then,

EVALUATE the constants A and B in the expansion

$$H(s) = \frac{4(s + 5)}{(s + 1)(s + 2)} \equiv \frac{A}{s + 1} + \frac{B}{s + 2}$$

5.4   Finally,

$H(s) = $ _____   (Partial Fraction form)

and so

$h(t) = $ _____   meter

CHECK your result at $t = 0$ against the I.C. given in frame **5.2**

To find A, first multiply by its denominator, $s + 1$:

$$\frac{4(s + 5)\cancel{(s + 1)}}{\cancel{(s + 1)}(s + 2)} = A + \frac{B(s + 1)}{(s + 2)}$$

Then, to eliminate the term in B, put $s = -1$

and

$$A = \left. \frac{4(s + 5)\cancel{(s + 1)}}{\cancel{(s + 1)}(s + 2)} \right|_{s = -i} = \frac{16}{1} = 16$$

Similarly,

$$B = \left. \frac{4(s + 5)\cancel{(s + 2)}}{(s + 1)\cancel{(s + 2)}} \right|_{s = -2} = \frac{12}{-1} = -12$$

Thus $\underline{A = 16}$ and $\underline{B = -12}$

$$H(s) = \underline{\frac{16}{s + 1} - \frac{12}{s + 2}}$$

$$h(t) = \underline{16\epsilon^{-t} - 12\epsilon^{-2t}} \text{ meter}$$

Check: $h(0) = 16 - 12 = 4$ which was the given I.C.
The graph of $h(t)$ follows:

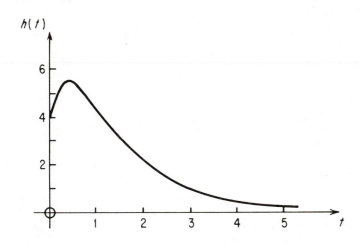

REVIEW

In each of the three systems problems we have just solved we followed the pattern proposed earlier. That is, we
1. *Transformed* the D.E. into an algebraic equation;
2. Solved this equation for the algebraic unknown; and
3. Took the inverse transform of the algebraic solution to find the corresponding unknown time function.

The Partial Fraction expansion should be regarded simply as an algebraic "trick" to put the algebraic solution into a form which matches entries in our table of transforms.

The three D.E.'s we have solved were each of first order.

Before we can transform higher order differential equations we must learn to transform the corresponding higher order derivatives.

We will now proceed from $\mathcal{L}[df/dt]$ to the transformation of derivatives of any order. *This next section will require 10–25 minutes.*

LAPLACE TRANSFORM OF THE SECOND DERIVATIVE

6.1 Although we could find $\mathcal{L}[d^2f(t)/dt^2]$ by evaluating the defining integral,

$$\mathcal{L}[d^2f(t)/dt^2] = \int_0^\infty \{d^2f(t)/dt^2\}\epsilon^{-st}\,dt,$$

we can avoid the integration by making use of the known result,

$$\mathcal{L}[df(t)/dt] = sF(s) - f(0) \qquad \text{(Theorem 2)}$$

We first define a new variable:

$$g(t) = \frac{df(t)}{dt} \qquad\qquad (1)$$

so that $\qquad \dfrac{d^2f(t)}{dt^2} = $ _____ $\qquad (2)$

6.2 Then, taking the L.T. of equation (1), with the help of Theorem 2:

$$G(s) = \text{_____} \qquad (3)$$

Similarly, using Theorem 2 to transform equation (2):

$$\mathcal{L}\left[\frac{d^2f(t)}{dt^2}\right] = \text{_____} \qquad (4)$$

6.3 As a matter of notation we write the initial value of $df(t)/dt$ as $df(t)/dt\,|_0$. That is,

$$g(0) = \left.\frac{df(t)}{dt}\right|_0 \qquad (5)$$

Finally, SUBSTITUTE (3) and (5) into equation (4) to obtain:

$$\mathcal{L}\left[\frac{d^2f(t)}{dt^2}\right] = \text{_____}$$

$$\frac{d^2 f(t)}{dt^2} = \frac{dg(t)}{dt} \qquad \text{differentiating equation (1)}$$

$$G(s) = sF(s) - f(0)$$

$$\mathcal{L}\left[\frac{d^2 f(t)}{dt^2}\right] = sG(s) - g(0)$$

$$\mathcal{L}\left[\frac{d^2 f(t)}{dt^2}\right] = s\{sF(s) - f(0)\} - \left.\frac{df(t)}{dt}\right|_0$$

or

$$\mathcal{L}\left[\frac{d^2 f(t)}{dt^2}\right] = s^2 F(s) - sf(0) - \left.\frac{df(t)}{dt}\right|_0$$

6.4   COPY the result

$$\mathcal{L}\left[\frac{d^2 f(t)}{dt^2}\right] = s^2 F(s) - sf(0) - \left.\frac{df(t)}{dt}\right|_0$$

onto line 14 of your semi-notes.
Then READ the comment on page 55, opposite.

LAPLACE TRANSFORM OF THE Nth DERIVATIVE

6.5   We could treat the L.T. of the third derivative in the same way as for the second, and we would find that by using the results

$$\mathcal{L}[df/dt] = sF(s) - f(0) \qquad \text{(Theorem 2)}$$

and $$\mathcal{L}[d^2 f/dt^2] = s^2 F(s) - sf(0) - df/dt\,|_0 \qquad \text{(Theorem 3)}$$

we would obtain

$$\mathcal{L}[d^3 f/dt^3] = s^3 F(s) - s^2 f(0) - sdf/dt\,|_0 - d^2 f/dt^2\,|_0$$

By inspection of the above, we would expect that

$$\mathcal{L}[d^n f/dt^n] = s^n F(s) - s^{n-1} f(0) - s^{n-2}\, df/dt\,|_0 - \cdots$$
$$\cdots - sd^{n-2} f/dt^{n-2}\,|_0 - d^{n-1} f/dt^{n-1}\,|_0$$

COPY this *carefully* onto line 15 of your semi-notes.

Comment:

From this point on we will often drop the explicit indication of the functional dependence on t (i.e., the independent variable). For example, we will write

$$f \quad \text{for} \quad f(t)$$

$$Ri \quad \text{for} \quad Ri(t)$$

$$dv/dt \quad \text{for} \quad dv(t)/dt$$

$$\int_0^t i\,dt \quad \text{for} \quad \int_0^t i(t)\,dt, \text{ etc.}$$

6.6 As an example,

$$\mathcal{L}\left[m\,\frac{d^2x}{dt^2}\right] = \underline{\hspace{5cm}}$$

6.7 Similarly,

$$\mathcal{L}\left[L\,\frac{d^2q}{dt^2}\right] = \underline{\hspace{5cm}}$$

6.8 Further,

$$\mathcal{L}\left[R\,\frac{dq}{dt}\right] = \underline{\hspace{3cm}},$$

$$\mathcal{L}\left[\frac{1}{C}\,q\right] = \underline{\hspace{3cm}},$$

C is a constant

$q(t)$ is a variable

and $\mathcal{L}[76] = \underline{\hspace{3cm}}$

6.9 Now, given the equation

$$L\,\frac{d^2q}{dt^2} + R\,\frac{dq}{dt} + \frac{1}{C}\,q = 76$$

with I.C.'s $q(0)$ and $dq/dt\,|_0$,
TRANSFORM term by term to obtain the *transformed* equation:

$$\underline{\hspace{8cm}} =$$

The next problem will require 15 to 30 minutes.

$$\mathcal{L}\left[m\frac{d^2x}{dt^2}\right] = m\;\mathcal{L}\left[\frac{d^2x}{dt^2}\right] = m\left\{s^2X(s) - sx(0) - \frac{dx}{dt}\bigg|_0\right\}$$

$$\mathcal{L}\left[L\frac{d^2q}{dt^2}\right] = L\;\mathcal{L}\left[\frac{d^2q}{dt^2}\right] = L\left\{s^2Q(s) - sq(0) - \frac{dq}{dt}\bigg|_0\right\}$$

$$\mathcal{L}\left[R\frac{dq}{dt}\right] = R\;\mathcal{L}\left[\frac{dq}{dt}\right] = R\{sQ(s) - q(0)\}$$

$$\mathcal{L}\left[\frac{1}{C}q\right] = \frac{1}{C}\;\mathcal{L}[q] = \frac{1}{C}Q(s) \qquad \text{(L.T. of a } \textit{constant times a variable})$$

$$\mathcal{L}[76] = 76/s \qquad\qquad \text{(L.T. of a } \textit{constant})$$

$$L\left\{s^2Q(s) - sq(0) - \frac{dq}{dt}\bigg|_0\right\} + R\{sQ(s) - q(0)\} + \frac{1}{C}Q(s) = 76/s$$

A PROBLEM IN MECHANICS

7.1   Here we will apply Theorem 3 to a simple situation in mechanics—a mass hanging on a spring, with a dashpot to provide viscous damping:

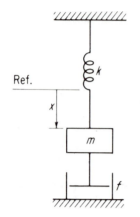

Ref.

The reference level is chosen so that the spring tension is zero when $x = 0$. Then, acting *downward* on the mass are:

$$\text{spring tension} = -kx$$
$$\text{damping force} = -f\,dx/dt$$
$$\text{gravitational force} = mg$$
$$\text{force of inertia} = -m\,d^2x/dt^2$$

Here f is the friction coefficient.

If we invoke Newton instead of Kirchhoff:

$$m\frac{d^2x}{dt^2} + f\frac{dx}{dt} + kx = mg \qquad \text{with I.C.'s } x(0) \text{ and } \frac{dx}{dt}\bigg|_0$$

then, **TRANSFORMING** term by term (remembering that mg = *constant*), we obtain the *transformed* equation

=

$$m\left\{s^2X(s) - sx(0) - \frac{dx}{dt}\bigg|_0\right\} + f\{sX(s) - x(0)\} + kX(s) = mg/s$$

transform of a
constant times a variable

transform of the *constant*, mg

7.2   In the equation

$$m\{s^2 X(s) - sx(0) - dx/dt\,|_0\} + f\{sX(s) - x(0)\} + kX(s) = mg/s$$

SUBSTITUTE the following data

$$m = 2\,\text{Kgm}, \quad f = 10\,\text{newton/m/sec}, \quad k = 12\,\text{n/m}, \quad g = 12\,\text{m/sec}^2,$$

$$x(0) = 9\,\text{m}, \quad \text{and} \quad dx/dt\,|_0 = -18\,\text{m/sec}.$$

to obtain

_____ $=$

(Don't manipulate the algebra.)

☐

7.3   SOLVE the equation above for $X(s)$:

Put your answer in the preferred form, $X(s) = \dfrac{(s^2 + \quad s + \quad)}{s(s^2 + \quad s + \quad)}$ ⎰— a constant

☐

$$2\{s^2 X(s) - 9s + 18\} + 10\{sX(s) - 9\} + 12X(s) = 24/s$$

If $\qquad 2\{s^2 X(s) - 9s + 18\} + 10\{sX(s) - 9\} + 12X(s) = 24/s$

then $\qquad 2\{s^2 + 5s + 6\}X(s) = 24/s + 18s - 36 + 90$

$$= \frac{18s^2 + 54s + 24}{s}$$

That is $\qquad X(s) = \dfrac{18s^2 + 54s + 24}{2s(s^2 + 5s + 6)} = \dfrac{9s^2 + 27s + 12}{s(s^2 + 5s + 6)}$

or $\qquad X(s) = \dfrac{9(s^2 + 3s + 4/3)}{s(s^2 + 5s + 6)}$

7.4   As part of the process of expanding into Partial Fractions we must factor the denominator. In this case it is easy to see that

$$X(s) = \frac{9(s^2 + 3s + 4/3)}{s(s^2 + 5s + 6)} = \frac{9s^2 + 27s + 12}{s(s + 2)(s + 3)} = \frac{A}{s} + \frac{B}{s + 2} + \frac{C}{s + 3}$$

Now FIND A, B, and C, using the method introduced earlier:

If
$$\frac{9s^2 + 27s + 12}{s(s + 2)(s + 3)} = \frac{A}{s} + \frac{B}{s + 2} + \frac{C}{s + 3}$$

then
$$A = \frac{(9s^2 + 27s + 12)s}{s(s + 2)(s + 3)}\bigg|_{s=0} = \frac{2}{}$$

(if we multiply by s and put $s = 0$ the last *two* terms go to zero.)

$$B = \frac{(9s^2 + 27s + 12)(s + 2)}{s(s + 2)(s + 3)}\bigg|_{s=-2} = \frac{36 - 54 + 12}{-2(1)} = \frac{3}{}$$

$$C = \frac{(9s^2 + 27s + 12)(s + 3)}{s(s + 2)(s + 3)}\bigg|_{s=-3} = \frac{81 - 81 + 12}{(-3)(-1)} = \frac{4}{}$$

(See frames **4.8** and **4.9** if you need to review the method leading to the above results.)

7.5   Now, knowing that

$$X(s) = \frac{2}{s} + \frac{3}{s+2} + \frac{4}{s+3},$$

$$x(t) = \underline{\hspace{5cm}} \text{ meter}$$

It follows that the velocity, dx/dt, is

$$\frac{dx}{dt} = \underline{\hspace{5cm}} \text{ meter/sec}$$

7.6   It is always wise to check the solution at least against the given I.C.'s.

Here, by substitution in the equations above,

$$x(0) = \underline{\hspace{3cm}} \text{ meter}$$

and $\left. \dfrac{dx}{dt} \right|_0 = \underline{\hspace{3cm}}$ m/sec   Do these values agree with those given in frame **7.2**?

7.7   Just as in the first problem, we can evaluate $x(t) = 2 + 3\epsilon^{-2t} + 4\epsilon^{-3t}$ for a series of values of t and then plot the results:

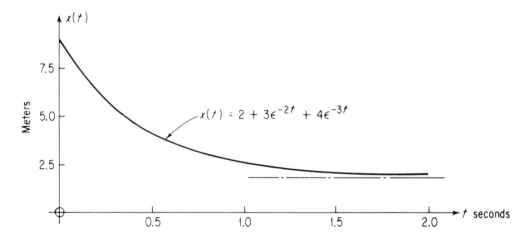

The graph suggests that the damper is relatively powerful—the mass does not oscillate on the end of the spring as we might expect, but "oozes" into equilibrium.

$$x(t) = \underline{2 + 3\epsilon^{-2t} + 4\epsilon^{-3t}} \text{ meter}$$

$$\frac{dx}{dt} = \underline{-6\epsilon^{-2t} - 12\epsilon^{-3t}} \text{ m/sec}$$

$$x(0) = \underline{9} \text{ meter} \qquad \text{putting } t = 0 \text{ in } x(t)$$

$$\left.\frac{dx}{dt}\right|_0 = \underline{-18} \text{ m/sec} \qquad \text{putting } t = 0 \text{ in } dx/dt$$

These clearly agree with the given I.C.'s.

LAPLACE TRANSFORM OF THE DEFINITE INTEGRAL

Now that we know how to treat higher derivatives it is logical that we should turn our attention to the L.T. of an integral, and, since integration is the "reverse" of differentiation, we might hope to find $\mathcal{L}\left[\int_0^t f(t)\,dt\right]$ from $\mathcal{L}[df(t)/dt]$.

8.1   We define a new variable, $g(t)$, whose *derivative* is equal to $f(t)$:

$$f(t) = \frac{dg(t)}{dt} \tag{1}$$

Then, integrating both sides,

$$\int_0^t f(t)\,dt = \int_0^t \frac{dg(t)}{dt}\,dt$$

$$=$$

← INTEGRATE the right hand side and then SUBSTITUTE the limits

That is, $\int_0^t f(t)\,dt = \underline{\hspace{4cm}}$ ←⎯⎯⎯ $\tag{2}$

8.2   We next TRANSFORM equation (2):

$$\mathcal{L}\left[\int_0^t f(t)\,dt\right] = \underline{\hspace{4cm}} \tag{3}$$

8.3   Finally, we can transform equation (1) to obtain

$$F(s) = sG(s) - g(0) \tag{4}$$

COMPARE equations (3) and (4), and COMPLETE the relation

$$\mathcal{L}\left[\int_0^t f(t)\,dt\right] = \underline{\hspace{3cm}} \quad \text{(in terms of } F(s) \text{ and } s)$$

$$\int_0^t f(t)\,dt \;=\; \int_0^t \frac{dg(t)}{dt}\,dt$$

$$=\; g(t)\Big|_0^t$$

— a constant!

$$=\; \underline{g(t) - g(0)}$$

$$\mathcal{L}\left[\int_0^t f(t)\,dt\right] \;=\; \underline{G(s) - g(0)/s}$$

— transform of the constant, $g(0)$

Since $F(s) = sG(s) - g(0)$

$G(s) - g(0)/s = F(s)/s$

Thus $$\boxed{\mathcal{L}\left[\int_0^t f(t)\,dt\right] = F(s)/s}$$

8.4   COPY the equation

$$\mathcal{L}\left[\int_0^t f(t)\, dt\right] = F(s)/s$$

onto line 16 of your semi-notes, as Theorem 5.

8.5   In summary, we see that the process of definite integration is transformed into the algebraic operation of _____ .

In contrast, we found earlier that the process of differentiation was transformed into the process of _____ , apart from the subtraction of the constant, $f(0)$.

8.6   Using Theorem 5,

$$\mathcal{L}\left[\frac{1}{C}\int_0^t i(t)\, dt\right] = \underline{\hspace{3cm}}$$

8.7   Similarly, given that $\mathcal{L}[A\epsilon^{at}] = A/(s-a)$,

$$\mathcal{L}\left[\int_0^t A\epsilon^{at}\, dt\right] = \underline{\hspace{3cm}}$$

8.8   And again, given that $\mathcal{L}[t\epsilon^{at}] = 1/(s-a)^2$,

$$\mathcal{L}\left[\int_0^t t\epsilon^{at}\, dt\right] = \underline{\hspace{3cm}} .$$

---

division or division by s

since $\mathcal{L}\left[\displaystyle\int_0^t f(t)\, dt\right] = F(s)/s$

multiplication or multiplication by s

since $\mathcal{L}\left[\dfrac{d}{dt} f(t)\right] = sF(s) - f(0)$

$\mathcal{L}\left[\dfrac{1}{C} \displaystyle\int_0^t i(t)\, dt\right] = \dfrac{1}{C}\,\dfrac{I(s)}{s}$

$\mathcal{L}\left[\displaystyle\int_0^t A\epsilon^{at}\, dt\right] = \dfrac{\mathcal{L}[A\epsilon^{at}]}{s} = \dfrac{A}{s(s-a)}$ In words: Divide the transform of the integrand by s.

$\mathcal{L}\left[\displaystyle\int_0^t t\epsilon^{at}\, dt\right] = \dfrac{\mathcal{L}[t\epsilon^{at}]}{s} = \dfrac{1}{s(s-a)^2}$

8.9 Given that $\mathcal{L}[A] = A/s$,

$$\mathcal{L}\left[\int_0^t A\, dt\right] = \underline{\hspace{4cm}}$$

8.10 Now, forget all about the L.T., and simply EVALUATE the following integral:

$$\int_0^t A\, dt = \underline{\hspace{4cm}}$$

8.11 Finally, knowing that $\mathcal{L}\left[\int_0^t A\, dt\right] = \underline{\hspace{2cm}}$ and that $\int_0^t A\, dt = \underline{\hspace{2cm}}$

it follows that $\mathcal{L}[At] = \underline{\hspace{2cm}}$

8.12 Given $\mathcal{L}[At]$ as in the previous frame,

$$\mathcal{L}\left[\int_0^t At\, dt\right] = \underline{\hspace{4cm}}$$

8.13 You can also EVALUATE the integral:

$$\int_0^t At\, dt = \underline{\hspace{4cm}}$$

8.14 Therefore, COMPARING frames **8.12** and **8.13**,

$$\mathcal{L}[At^2/2] = \underline{\hspace{4cm}} \quad \text{(a function of } s\text{)}$$

$$\mathcal{L}\left[\int_0^t A \, dt\right] = \frac{A/s}{s} = \underline{A/s^2}$$

$$\int_0^t A \, dt = At \Big|_0^t = \underline{At}$$

$$\mathcal{L}\left[\int_0^t A \, dt\right] = \underline{A/s^2} \quad \text{and} \quad \int_0^t A \, dt = \underline{At}$$

$$\therefore \mathcal{L}[At] = \underline{A/s^2}$$

From the previous frame, $\mathcal{L}[At] = A/s^2$

therefore, $\quad \mathcal{L}\left[\int_0^t At \, dt\right] = \frac{A/s^2}{s} = \underline{A/s^3}$

$$\int_0^t At \, dt = At^2/2 \Big|_0^t = \underline{At^2/2}$$

$$\mathcal{L}[At^2/2] = \mathcal{L}\left[\int_0^t At \, dt\right] = \underline{A/s^3}$$

8.15   Similarly, given that $\mathcal{L}\left[\dfrac{At^2}{2}\right] = A/s^3$,

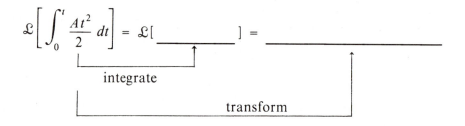

$$\mathcal{L}\left[\int_0^t \frac{At^2}{2}\,dt\right] = \mathcal{L}[\underline{\hspace{1.5cm}}] = \underline{\hspace{4cm}}$$

integrate

transform

□

8.16   To "flog a dead horse,"

$$\mathcal{L}\left[\int_0^t \frac{At^3}{6}\,dt\right] = \mathcal{L}[\underline{\hspace{1.5cm}}] = \underline{\hspace{3cm}}$$

□

8.17   Summarizing, we have

$$\mathcal{L}[A] = \mathcal{L}[At^0] = A/s$$

$$\mathcal{L}[At^1] = A/s^2$$

$$\mathcal{L}[At^2/2] = A/s^3$$

$$\mathcal{L}[At^3/6] = \underline{\hspace{2cm}}$$

$$\mathcal{L}[At^4/24] = \underline{\hspace{2cm}}$$

and we would expect that

$$\mathcal{L}[At^n/\underline{\hspace{1.5cm}}] = \underline{\hspace{3cm}}$$

□

$$\mathcal{L}\left[\int_0^t \frac{At^2}{2}\, dt\right] = \mathcal{L}\left[\frac{At^3}{6}\right] = \frac{A/s^3}{s} = \underline{A/s^4}$$

overbrace: integrate

Theorem 5

$$\mathcal{L}\left[\int_0^t \frac{At^3}{6}\, dt\right] = \mathcal{L}\left[\frac{At^4}{24}\right] = \frac{A/s^4}{s} = \underline{A/s^5}$$

$$\mathcal{L}[At^3/6] = \underline{A/s^4}$$

$$\mathcal{L}[At^4/24] = \underline{A/s^5}$$

$$\mathcal{L}[At^n/\underline{n!}] = \underline{A/s^{n+1}} \qquad (n! = n \text{ factorial})$$

8.18   COPY the transform pair

$$At \leftrightarrow A/s^2$$

onto line 7 of your semi-notes.

8.19   We showed that

$$\mathcal{L}[At^n/n!] = A/s^{n+1}$$

Therefore,

$$\boxed{\mathcal{L}[At^n] = An!/s^{n+1}}$$ (multiplying both sides by $n!$)

COPY the transform pair

$$At^n \leftrightarrow An!/s^{n+1}$$ n = positive integer

onto line 8 of your semi-notes.

The next problem will require 20 to 40 minutes.

SOLUTION OF CIRCUIT PROBLEM III

9.1   Here is a circuit problem requiring the application of our newly acquired skills. We are to find $i(t)$ for $t \geq 0$.

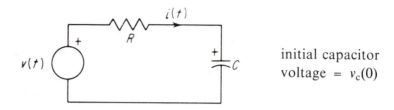

initial capacitor
voltage $= v_c(0)$

The capacitor voltage is

$$v_c = \frac{1}{C} \int_0^t i\, dt + v_c(0)$$

and therefore, from Kirchhoff's voltage law:

$$Ri + \frac{1}{C} \int_0^t i\, dt + v_c(0) = v(t)$$

9.2   TRANSFORMING,

$$\mathcal{L}[Ri(t)] = \underline{\hspace{3cm}}$$

$$\mathcal{L}\left[\frac{1}{C} \int_0^t i(t)\, dt \right] = \underline{\hspace{3cm}}$$

$$\mathcal{L}[v_c(0)] = \underline{\hspace{3cm}}$$

9.3   Therefore, the *transformed* equation is:

$$\underline{\hspace{6cm}} = \underline{\hspace{4cm}}$$

$$\mathcal{L}[Ri(t)] = \underline{RI(s)}$$

$$\mathcal{L}\left[\frac{1}{C}\int_0^t i(t)\,dt\right] = \underline{I(s)/Cs}$$

$$\mathcal{L}[v_c(0)] = \underline{v_c(0)/s} \qquad v_c(0) \text{ is a constant.}$$

$$\underline{RI(s) + \frac{I(s)}{Cs} + \frac{v_c(0)}{s} = V(s)}$$

9.4 In this problem we are given that

$$v(t) = 2t \quad \text{so that} \quad V(s) = \underline{\hspace{2cm}}$$

We are also given,

$$R = 1\,\Omega, \quad C = \tfrac{1}{2}\,\text{Farad}, \quad \text{and} \quad v_c(0) = 1\,\text{volt}.$$

SUBSTITUTE this data in the transformed equation

$$RI(s) + I(s)/Cs + v_c(0)/s = V(s)$$

and SOLVE for $I(s)$:

9.5 EXPRESS the above solution for $I(s)$ in Partial Fractions. Then, CHECK your result.

If $v(t) = 2t$, $V(s) = \underline{2/s^2}$

(line 7 of semi-notes)

Comment:
$v(t) = 2t$ is a voltage "ramp"
function of slope 2.

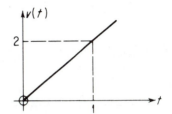

$$\left\{ 1 + \frac{2}{s} \right\} I(s) = \frac{2}{s^2} - \frac{1}{s}$$

$$\frac{s + 2}{s} I(s) = \frac{2 - s}{s^2}$$

$$I(s) = -\frac{s - 2}{s(s + 2)} \qquad \text{(or equivalent)}$$

If

$$I(s) = -\frac{s - 2}{s(s + 2)} = \frac{A}{s} + \frac{B}{s + 2}$$

then

$$A = -\frac{(s - 2)\cancel{s}}{\cancel{s}(s + 2)}\bigg|_{s=0} = 1$$

and

$$B = -\frac{(s - 2)\cancel{(s + 2)}}{s\cancel{(s + 2)}}\bigg|_{s=-2} = -2$$

Therefore

$$I(s) = \frac{1}{s} - \frac{2}{s + 2}$$

Check:

$$\frac{1}{s} - \frac{2}{s + 2} = \frac{s + 2 - 2s}{s(s + 2)} = -\frac{s - 2}{s(s + 2)}$$

9.6   With $I(s) = 1/s - 2/(s + 2)$,

$i(t) =$ _____

$\Box$

A REVIEW PROBLEM

9.7   As review we will now solve a D.E. (strictly, an integro-differential equation) containing a derivative and an integral of the variable, $y(t)$.

TRANSFORM the equation

$$2\frac{dy}{dt} + 6y + 4\int_0^t y\,dt = 20\epsilon^{-3t} \qquad \text{with I.C. } y(0)$$

to obtain the *transformed* equation:

_____ $=$ _____

$\Box$

9.8   Given that $y(0) = 0$, SOLVE the transformed equation for $Y(s)$:

$\Box$

$i(t) = \underline{1 - 2\epsilon^{-2t}}$ amp

—transform of $6y(t)$

$$\underline{2\{sY(s) - y(0)\} + 6\overset{\downarrow}{Y}(s) + 4Y(s)/s = 20/(s + 3)}$$

(or equivalent)

$\{2s + 6 + 4/s\}Y(s) = 20/(s + 3)$

$\{s^2 + 3s + 2\}Y(s) = 10s/(s + 3)$ multiplying by $s/2$

$$Y(s) = \underline{\frac{10s}{(s^2 + 3s + 2)(s + 3)}} \quad \text{or} \quad \underline{\frac{10s}{(s + 1)(s + 2)(s + 3)}}$$

9.9   EXPRESS $Y(s) = \dfrac{10s}{(s + 1)(s + 2)(s + 3)}$ in Partial Fractions

and then CHECK your result:

⬜

_____

9.10   Finally, $y(t) =$ _____

⬜

_____

If $\quad \dfrac{10s}{(s+1)(s+2)(s+3)} = \dfrac{A}{s+1} + \dfrac{B}{s+2} + \dfrac{C}{s+3}$

then $\qquad A = \left.\dfrac{10s\cancel{(s+1)}}{\cancel{(s+1)}(s+2)(s+3)}\right|_{s=-1} = \dfrac{-10}{2} = -5$

$\qquad\qquad B = \left.\dfrac{10s\cancel{(s+2)}}{(s+1)\cancel{(s+2)}(s+3)}\right|_{s=-2} = \dfrac{-20}{-1} = 20$

$\qquad\qquad C = \left.\dfrac{10s\cancel{(s+3)}}{(s+1)(s+2)\cancel{(s+3)}}\right|_{s=-3} = \dfrac{-30}{2} = -15$

Thus $\quad Y(s) = \dfrac{-5}{s+1} + \dfrac{20}{s+2} - \dfrac{15}{s+3}$

Check: $\quad Y(s) = \dfrac{-5(s+2)(s+3) + 20(s+1)(s+3) - 15(s+1)(s+2)}{(s+1)(s+2)(s+3)}$

$\qquad\qquad\quad = \dfrac{0s^2 + 10s + 0}{(s+1)(s+2)(s+3)} \qquad$ which checks.

$y(t) = -5\epsilon^{-t} + 20\epsilon^{-2t} - 15\epsilon^{-3t}$

This section should take you between 30 and 60 minutes.

SOME REVIEW ITEMS

10.1   *Without* looking at your semi-notes *or* at earlier frames, WRITE DOWN the *definition* of the L.T. of $f(t)$:

$$\mathcal{L}[f(t)] =$$

10.2   *Without* looking back at earlier frames, *or* at your semi-notes, *use the defining integral* above to EVALUATE $\mathcal{L}[A\epsilon^{at}]$. (Assume that *no* transform pairs are known.)

10.3   Use the transform pair that you have just derived to FIND

$$\mathcal{L}[K\epsilon^{j\omega t}] = \underline{\hspace{4cm}}$$

where K and ω are constants, and $j = \sqrt{-1}$

$$\mathcal{L}[f(t)] \;=\; \int_0^\infty f(t)\,\epsilon^{-st}\,dt$$

$$\mathcal{L}[A\epsilon^{at}] \;=\; A\int_0^\infty \epsilon^{at}\,\epsilon^{-st}\,dt$$

$$=\; A\int_0^\infty \epsilon^{(a-s)t}\,dt$$

$$=\; \frac{A\epsilon^{(a-s)t}}{a-s}\Bigg|_0^\infty$$

$$=\; \frac{A}{a-s}\{\epsilon^{(a-s)\infty} - \epsilon^{(a-s)0}\}$$

$$\mathcal{L}[A\epsilon^{at}] \;=\; \frac{A}{s-a} \qquad \text{(assuming that } (a-s)<0, \text{ or more strictly, that } \mathfrak{Re}(a-s)<0\text{)}$$

(This was our first transform pair—line 5 of semi-notes)

$$\mathcal{L}[K\epsilon^{j\omega t}] \;=\; K/(s-j\omega) \qquad \text{(the result } \mathcal{L}[A\epsilon^{at}] = A/(s-a)\text{ is valid for complex } a \text{ provided that } \mathfrak{Re}(a-s)<0\text{)}$$

10.4 To find $\mathcal{L}[A \cos \omega t]$, where A *and* ω *are constants*, we could evaluate the defining integral

$$\mathcal{L}[A \cos \omega t] = \int_0^\infty A \cos \omega t \, \epsilon^{-st} \, dt.$$

However, it is simpler to express the cosine in exponential form,

$$A \cos \omega t = \frac{A}{2} \{\epsilon^{j\omega t} + \epsilon^{-j\omega t}\},$$

since we already know the L.T. of an exponential (line 5 of semi-notes).

Thus, $\quad \mathcal{L}[A \cos \omega t] = \mathcal{L}\left[\frac{A}{2} \{\epsilon^{j\omega t} + \epsilon^{-j\omega t}\}\right]$

$$=$$

$$=$$

$$=$$

$$= \underline{\qquad} /(\underline{\qquad})$$

(The answer should be real.)

10.5 Using the *result* of the last frame,

$$\mathcal{L}[10 \cos 100t] = \underline{\qquad\qquad\qquad}$$

Comment:

From the Euler relation, $\epsilon^{j\theta} = \cos\theta + j\sin\theta$, it can be shown that $\cos\theta = \frac{1}{2}\{\epsilon^{j\theta} + \epsilon^{-j\theta}\}$. In this problem $\theta = \omega t$.

$$\mathcal{L}[A\cos\omega t] = \frac{A}{2}\{\mathcal{L}[\epsilon^{j\omega t}] + \mathcal{L}[\epsilon^{-j\omega t}]\}$$

$$= \frac{A}{2}\{1/(s - j\omega) + 1/(s + j\omega)\}$$

$$= \frac{A}{2}\left\{\frac{s + j\omega + s - j\omega}{(s - j\omega)(s + j\omega)}\right\}$$

$$\boxed{\mathcal{L}[A\cos\omega t] = As/(s^2 + \omega^2)}$$

$$\mathcal{L}[10\cos 100t] = \underline{10s/(s^2 + 100^2)}$$

10.6   COPY the transform pair

$$A \cos \omega t \longleftrightarrow As/(s^2 + \omega^2)$$

onto line 9 of your semi-notes.

10.7   The L.T. of the sine may be similarly evaluated. Thus

$$\mathcal{L}[A \sin \omega t] = \mathcal{L}\left[\frac{A}{2j} \{\epsilon^{j\omega t} - \epsilon^{-j\omega t}\}\right]$$

$$=$$

$$= \underline{\qquad}/(\qquad)$$ (Again the answer should be *real*.)

□

10.8   Alternatively, we can find $\mathcal{L}[A \sin \omega t]$ from $\mathcal{L}[A \cos \omega t]$, which is now known. Thus simple calculus allows us to relate the sine and cosine by

$$A \sin \omega t = \int_0^t A\omega \cos \omega t \, dt.$$

and, if we transform both sides of the equation, using the results on lines 9 and 16 of the semi-notes:

$$\mathcal{L}[A \sin \omega t] = \mathcal{L}\left[\int_0^t A\omega \cos \omega t \, dt\right]$$ (A and ω are constants)

$$=$$

$$= \underline{\qquad}/(\qquad)$$

Try to get your answers to this and the last frame to agree *before* looking at the answer, opposite.

□

Check your result by completing the alternative evaluation in the next frame (**10.8**).

You should have found that

$$\mathcal{L}[A \sin \omega t] = A\omega/(s^2 + \omega^2)$$

If you don't agree, in one or both frames, check your work. The solution to the problem on frame **10.7** follows exactly the same pattern as that of frame **10.4**, while frame **10.8** merely requires the direct application of Theorem 5, with $f(t) = A\omega \cos \omega t$.

10.9   COPY the transform pair

$$A \sin \omega t \longleftrightarrow A\omega/(s^2 + \omega^2)$$

onto line 10 of your semi-notes.

10.10   Using the above transform pair,

$$\mathcal{L}[10 \sin 100t] = \underline{\hspace{5cm}}$$

10.11   We can use the trigonometric identity

$$\cos(\alpha + \beta) = \cos \alpha \cos \beta - \sin \alpha \sin \beta$$

to EVALUATE

$$\mathcal{L}[A \cos(\omega t + \theta)] = \mathcal{L}[A\{\cos \omega t \cos \theta - \sin \omega t \sin \theta\}]$$

$$= \qquad\qquad\qquad\qquad\qquad\qquad \text{(note that } A, \omega, \theta, \cos \theta, \text{ and}$$
$$\sin \theta \text{ are } constants)$$

$$=$$

$$=$$

$$= \underline{\hspace{5cm}}$$

Put your answer in the preferred form: $\dfrac{\text{constant}(s \qquad\quad)}{s^2 + \omega^2}$

90

$$\mathcal{L}\,[10 \sin 100t] \;=\; \underline{1000/(s^2 + 100^2)}$$

$$\mathcal{L}\,[A \cos(\omega t + \theta)] \;=\; \mathcal{L}\,[A\{\underline{\cos \omega t \cos \theta} \;-\; \underline{\sin \omega t \sin \theta}\}]$$

$$= A\left\{\cos\theta \,\frac{s}{s^2 + \omega^2} \;-\; \sin\theta \,\frac{\omega}{s^2 + \omega^2}\right\}$$

$$= A\left\{\frac{s \cos\theta \;-\; \omega \sin\theta}{s^2 + \omega^2}\right\}$$

$$= \underline{\frac{A \cos\theta \,(s \;-\; \omega \tan\theta)}{s^2 + \omega^2}}$$

THE FINAL TRANSFORM PAIR

Masses on springs, pendula, and L-R-C circuits often oscillate with amplitudes which decay exponentially with time. For this reason it behooves us to find the L.T. of the function

$$f(t) = M\epsilon^{\alpha t} \cos(\omega t + \theta)$$

where M, α, ω, and θ are real *constants* describing the damped oscillation. (α must be negative if the amplitude of the oscillation, $M\epsilon^{\alpha t}$, is decaying.) See sketch opposite.

10.12 Once again we can avoid direct evaluation of the defining integral by expressing the cosine in exponential form. Thus

$$M\epsilon^{\alpha t} \cos(\omega t + \theta) = M\epsilon^{\alpha t} \frac{\epsilon^{j(\omega t + \theta)} + \epsilon^{-j(\omega t + \theta)}}{2}$$

$$= \frac{M}{2}\{\epsilon^{\alpha t}\epsilon^{j\omega t}\epsilon^{j\theta} + \epsilon^{\alpha t}\epsilon^{-j\omega t}\epsilon^{-j\theta}\}$$

But $\epsilon^{j\theta}$ and $\epsilon^{-j\theta}$ are *constants*, which can be collected together with the $M/2$,

$$\therefore M\epsilon^{\alpha t} \cos(\omega t + \theta) = \tfrac{1}{2}M\epsilon^{j\theta}\epsilon^{(\alpha + j\omega)t} + \tfrac{1}{2}M\epsilon^{-j\theta}\epsilon^{(\alpha - j\omega)t}$$

Now you can use the transform pair for the exponential (line 5 of semi-notes) to TRANSFORM the last equation, term by term:

$$\mathcal{L}[M\epsilon^{\alpha t} \cos(\omega t + \theta)] = \underline{\hspace{3cm}} + \underline{\hspace{3cm}}$$

(*Don't manipulate the algebra.*)

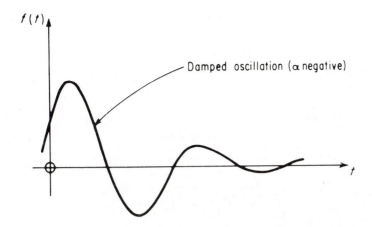

$$\frac{\frac{1}{2}M\epsilon^{j\theta}}{s-(\alpha+j\omega)}+\frac{\frac{1}{2}M\epsilon^{-j\theta}}{s-(\alpha-j\omega)} \quad \text{or} \quad \frac{\frac{1}{2}M\epsilon^{j\theta}}{s-\alpha-j\omega}+\frac{\frac{1}{2}M\epsilon^{-j\theta}}{s-\alpha+j\omega}$$

($\frac{1}{2}M\epsilon^{j\theta}$ and $\frac{1}{2}M\epsilon^{-j\theta}$ are constants.)

10.13   The result just reached is a little tidier when multiplied through by 2, yielding the transform pair:

$$2 M \epsilon^{\alpha t} \cos(\omega t + \theta) \longleftrightarrow \frac{M \epsilon^{j\theta}}{s - \alpha - j\omega} + \frac{M \epsilon^{-j\theta}}{s - \alpha + j\omega}$$

COPY this pair *very* carefully onto line 11 of your semi-notes.
CHECK your entry, especially with respect to signs.

10.14   Using the transform pair above

$$\mathcal{L}[10\epsilon^{-6t} \cos(10t + 0.5)] = \qquad +$$

10.15   Similarly,

$$\mathcal{L}[50\epsilon^{2t} \cos(377t - 0.25)] = \qquad +$$

10.16   *Using the same transform pair,*

$$\mathcal{L}[10 \cos 100t] = \qquad +$$

$$= \qquad \qquad \text{(combine the two terms)}$$

10.17   If $\quad Q(s) = \dfrac{12\epsilon^{j0.3}}{s + 7 - j8} + \dfrac{12\epsilon^{-j0.3}}{s + 7 + j8}$

then $\quad q(t) = $ _____   (damped cosine form)

$$\mathcal{L}[10\epsilon^{-6t}\cos(10t + 0.5)] = \underline{\frac{5\epsilon^{j0.5}}{s + 6 - j10} + \frac{5\epsilon^{-j0.5}}{s + 6 + j10}} \qquad \begin{aligned} M &= 5 \\ \alpha &= -6 \\ \omega &= 10 \\ \theta &= 0.5 \end{aligned}$$

CHECK the signs carefully!

$$\mathcal{L}[50\epsilon^{2t}\cos(377t - 0.25)] = \underline{\frac{25\epsilon^{-j0.25}}{s - 2 - j377} + \frac{25\epsilon^{j0.25}}{s - 2 + j377}}$$

CHECK your answer against that of frame **10.5**

(here $\alpha = 0$ and therefore $\epsilon^{\alpha t} = 1$)

$$q(t) = \underline{24\epsilon^{-7t}\cos(8t + 0.3)} \qquad (M = 12)$$

You will need 20–40 minutes to solve this problem.

RESPONSE OF A SERIES L-R-C CIRCUIT TO A RAMP INPUT

11.1 The aim is to find $i(t)$ in the following circuit for $t \geq 0$.

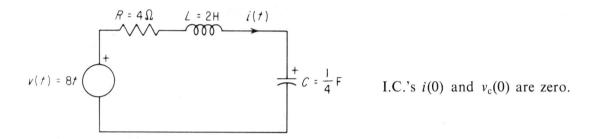

I.C.'s $i(0)$ and $v_c(0)$ are zero.

The D.E. is

$$L \frac{di}{dt} + Ri + \frac{1}{C} \int_0^t i\, dt + v_c(0) = v(t)$$

and the *transformed* equation is therefore:

(*Don't substitute data.*)

11.2 SUBSTITUTE the data in the transformed equation, and SOLVE for $I(s)$:

$$L\{sI(s) - i(\overset{=\,0}{\cancel{0}})\} + RI(s) + \frac{I(s)}{Cs} + \frac{v_c(\overset{=\,0}{\cancel{0}})}{\cancel{s}} = V(s) \quad \text{or} \quad \frac{8}{s^2}$$

(Your answer is acceptable if you omitted $i(0)$ and $v_c(0)/s$, since they are zero in this problem.)

$$2sI(s) + 4I(s) + 4I(s)/s = 8/s^2$$

$$2\{s + 2 + 2/s\}I(s) = 8/s^2$$

$$I(s) = \frac{4}{s(s^2 + 2s + 2)} \qquad \text{or equivalent}$$

11.3   Before we can expand $\dfrac{4}{s(s^2 + 2s + 2)}$ into Partial Fractions we must find the roots of the quadratic.

That is, FIND the *roots* of $s^2 + 2s + 2 = 0$:

11.4   It follows from the previous frame that $s^2 + 2s + 2$ can be FACTORED as below:

$$s^2 + 2s + 2 = (\underline{\hspace{3cm}})(\underline{\hspace{3cm}})$$

Using the quadratic formula, these roots are

$$\tfrac{1}{2}\{-2 + \sqrt{2^2 - 4(2)}\} \quad \text{and} \quad \tfrac{1}{2}\{-2 - \sqrt{2^2 - 4(2)}\}$$

or $\quad \underline{-1 + j} \quad$ and $\quad \underline{-1 - j}$

$$s^2 + 2s + 2 = \underline{(s + 1 - j1)(s + 1 + j1)} \quad \text{or, of course}$$

$$\underline{(s + 1 + j1)(s + 1 - j1)}$$

Note how the *roots* and *factors* are related. As a check, if you multiply the two factors together you should get $s^2 + 2s + 2$. Do you?

11.5   At this point we can write:

$$I(s) = \frac{4}{s(s^2 + 2s + 2)}$$

$$= \frac{4}{s(s + 1 - j)(s + 1 + j)} = \frac{A}{s} + \frac{B}{s + 1 - j} + \frac{C}{s + 1 + j}$$

CAUTION!

There is a trap here for the unwary. If we are to match the entry on line 11 of the semi-notes, the *order* of the last two terms is important. The first denominator should contain "$-j\omega$," the second should contain "$+j\omega$." Here, where $\omega = 1$, the proper order has been maintained. Careless reversal of the order will lead to an error in the sign of θ.

Now, FIND A in the normal way:

11.6   The constant, B, will turn out to be a complex number, but you can calculate it in the usual way. Therefore, FIND B, expressing your answer *in the rectangular form, $\alpha + j\beta$:*

$$A = \left. \frac{4\cancel{s}}{\cancel{s}(s^2 + 2s + 2)} \right|_{s=0} = \frac{4}{2} = \underline{2}$$

$$B = \left. \frac{4(\cancel{s+1-j})}{s(\cancel{s+1-j})(s + 1 + j)} \right|_{s=-1+j} = \frac{4}{(-1 + j)2j} = \frac{2}{-1 - j}$$

Rationalizing:

$$B = \frac{2(-1 + j)}{(-1 - j)(-1 + j)} = \frac{2(-1 + j)}{2} = \underline{\underline{-1 + j}}$$

11.7 MARK the complex value of $B = -1 + j$ on the diagram. *Then*, EXPRESS the value of B in polar (exponential) form, $M\epsilon^{j\theta}$:

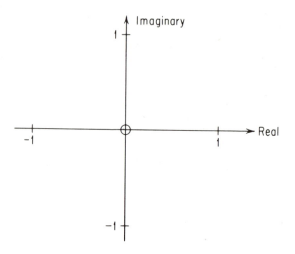

(Give θ in degrees.)

$$\boxed{}$$

11.8 Now compare

$$B = \left.\frac{4(s + 1 - j)}{s(s + 1 - j)(s + 1 + j)}\right|_{s = -1+j}$$

with

$$C = \left.\frac{4(s + 1 + j)}{s(s + 1 - j)(s + 1 + j)}\right|_{s = -1-j}$$

Note that these quantities differ only by the relative reversal of the sign in front of each j.

Therefore, *B and C are complex conjugates*, and thus

$$C = \underline{\hspace{4cm}} \quad \text{(polar form)}$$

$$\boxed{}$$

Given $B = -1 + j$

$M = \sqrt{1^2 + 1^2} = \sqrt{2}$ or 1.414

$\theta = 90° + \tan^{-1}\dfrac{1}{1}$ (see diagram)

$\qquad = 90° + 45°$

$\qquad = 135°$

$\therefore B = 1.414\,\epsilon^{j135°}$

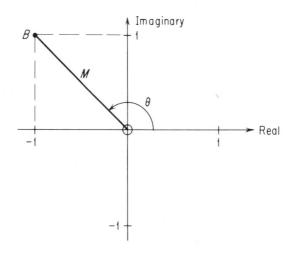

Comment:
Strictly speaking, all angles should be given in radian measure. However, it is common (if poor) practice to use degrees, largely because tables and slide-rules are so constructed.

$C = 1.414\,\epsilon^{-j135°}$ \qquad (complex conjugate of $B = 1.414\,\epsilon^{j135°}$)

Summarizing,

$$I(s) = \frac{2}{s} + \frac{1.414\epsilon^{j135°}}{s + 1 - j} + \frac{1.414\epsilon^{-j135°}}{s + 1 + j}$$

Therefore, using your table of transform pairs (lines 6 and 11),

$$i(t) = \underline{\hspace{7cm}} \quad amp$$

CHECK your answer at $t = 0$:

11.10   As in the first problem we solved, we can evaluate the expression for $i(t)$ for a series of values of t, and then plot the results. Notice that we must be careful to express the two angles in the argument of the cosine in the *same* units, *either* radians *or* degrees.

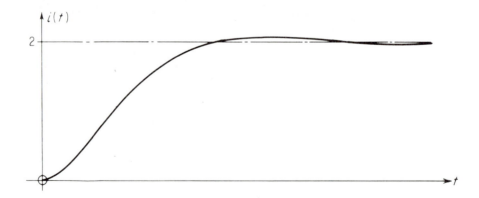

The oscillatory part of the solution decays quite rapidly, leaving $i(t) \rightarrow 2$ as $t \rightarrow \infty$.

$$i(t) = \underline{2 + 2.828\,\epsilon^{-t}\cos(t + 135°)}\ \text{amp}$$

Check: $i(0) = 2 + 2.828\cos 135° = 2 - 2 = 0$
which was the value given for the I.C. in frame **11.1**.

Comment:
Note that the two complex terms in $I(s)$ *are* in the same order as the two corresponding terms on line 11 of the semi-notes.

This problem can be solved in 15–30 minutes

AIRPLANE ROLL RESPONSE

12.1   Here we will solve for the rate of roll, $p(t)$, of an airplane whose ailerons are oscillated sinusoidally:

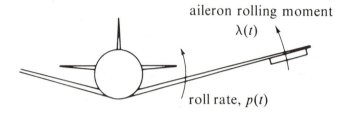

aileron rolling moment
$\lambda(t)$

roll rate, $p(t)$

Inertia torque $= J\dfrac{dp(t)}{dt}$

Damping torque $= f p(t)$

Aileron torque $= \lambda(t)$

Initial roll rate $= p(0)$

Summing the moments about the roll axis:

$$J\,\frac{dp(t)}{dt} + f p(t) = \lambda(t) \qquad J \text{ and } f \text{ are constants}$$

whence the *transformed* equation is

$$\underline{\hspace{8cm}} = \Lambda(s)$$

Notation: $\mathcal{L}[\lambda(t)] = \Lambda(s)$

12.2   SUBSTITUTE the data

$$J = 0.5, \quad f = 1.5, \quad p(0) = 1, \quad \text{and} \quad \lambda(t) = \cos t$$

and SOLVE for $P(s)$:

(*Don't* factor the quadratic term.)

$$J\{sP(s) - p(0)\} + fP(s) = \Lambda(s)$$

Comment:
Λ and λ are upper and lower case "lambda."

Substituting directly, with $\Lambda(s) = \mathcal{L}[\cos t] = s/(s^2 + 1^2)$:

$$\tfrac{1}{2}\{sP(s) - 1\} + \tfrac{3}{2}P(s) = s/(s^2 + 1^2)$$

or $\quad \tfrac{1}{2}(s + 3)P(s) = \tfrac{1}{2} + s/(s^2 + 1^2) = \dfrac{s^2 + 2s + 1}{2(s^2 + 1^2)}$

$$\therefore P(s) = \dfrac{s^2 + 2s + 1}{(s^2 + 1^2)(s + 3)} \qquad \text{(or equivalent)}$$

12.3 We proceed as usual to the Partial Fraction expansion (P.F.E.), always factoring the denominator into first order terms. Thus

$$P(s) = \frac{s^2 + 2s + 1}{(s^2 + 1^2)(s + 3)} = \frac{s^2 + 2s + 1}{(s - j)(s + j)(s + 3)} = \frac{A}{s - j} + \frac{B}{s + j} + \frac{C}{s + 3}$$

Now FIND A in the normal way. Express your answer in the rectangular form, $\alpha + j\beta$:

☐

12.4 MARK the value of A on the diagram. *Then*, EXPRESS the value of A in exponential (polar) form, $M\epsilon^{j\theta}$:

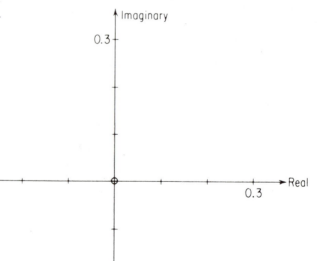

(*Data*: $\tan^{-1} 0.333 \doteq 18°.4$ and $\sqrt{0.1} \doteq 0.316$)

☐

$$A = \frac{(s^2 + 2s + 1)(\cancel{s - j})}{(\cancel{s - j})(s + j)(s + 3)}\Bigg|_{s=j}$$

$$= \frac{-1 + 2j + 1}{(2j)(j + 3)} = \frac{2j}{(2j)(3 + j)} = \frac{1}{3 + j}$$

$$= \frac{(3 - j)}{(3 + j)(3 - j)} = \frac{3 - j}{9 + 1} \quad \text{(rationalizing)}$$

or $\quad \underline{A = 0.3 - j0.1}$

Since $\quad A = 0.3 - j0.1$

$$M = \sqrt{0.3^2 + 0.1^2} = \sqrt{0.1} = 0.316$$

and $\quad \theta = \tan^{-1}(-0.1/0.3)$

$$= -\tan^{-1}0.333 = -18°.4$$

or $\quad A = 0.3 - j0.1 = \underline{0.316\epsilon^{-j18°.4}}$

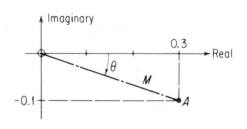

12.5 Returning to $P(s) = \dfrac{s^2 + 2s + 1}{(s - j)(s + j)(s + 3)} = \dfrac{A}{s - j} + \dfrac{B}{s + j} + \dfrac{C}{s + 3}$

we have already $A = \dfrac{(s^2 + 2s + 1)(s - j)}{(s - j)(s + j)(s + 3)}\Bigg|_{s = j}$

and similarly, $B = $ \rule{3cm}{0.4pt} (in the same form)

12.6 Now COMPARE the expressions for A and B above. As a result of the comparison, and knowing that $A = 0.316\epsilon^{-j18°.4}$ you can write

$B = $ \rule{4cm}{0.4pt} (polar form)

12.7 To save you some calculation we have

$$C = \frac{(s^2 + 2s + 1)(s + 3)}{(s - j)(s + j)(s + 3)}\Bigg|_{s = -3} = \frac{9 - 6 + 1}{(-3 - j)(-3 + j)} = 0.4$$

Therefore, in summary:

$$P(s) = \frac{s^2 + 2s + 1}{(s - j)(s + j)(s + 3)} = \frac{0.316\epsilon^{-j18°.4}}{s - j} + \frac{0.316\epsilon^{j18°.4}}{s + j} + \frac{0.4}{s + 3}$$

From which it follows that

$p(t) = $ \rule{6cm}{0.4pt}

(Use the transform pair on line 11 of the semi-notes for the first two terms and watch the signs!)

$$B = \left. \frac{(s^2 + 2s + 1)(\cancel{s + j})}{(s - j)(\cancel{s + j})(s + 3)} \right|_{s = -j}$$

$$B = \underline{0.316\epsilon^{j18°.4}} \qquad (A \text{ and } B \text{ are complex conjugates.})$$

$$p(t) = \underline{0.632 \cos (t - 18°.4) + 0.4\epsilon^{-3t}}$$

Comments:
1. The first two terms in the Partial Fraction expansion *are* in the correct order, matching the entry on line 11 of the semi-notes.
2. The constant, α, in the same entry, is here zero.

12.8   CHECK your solution

$$p(t) = 0.632 \cos(t - 18°.4) + 0.4\epsilon^{-3t}$$

against the given initial condition, $p(0) = 1$:

SUMMARY:

The last two problems gave rise to complex terms in the Partial Fraction expansion. However, the basic technique was not changed. We have only to remember that:
1. The complex constants in the P.F.E. must be expressed in polar form to match the entry on line 11 of the semi-notes;
2. The *order* of the terms must match that entry (see comment on frame **11.5**); and
3. The complex constants appear as complex conjugates.

A further example, which will take 15 to 30 minutes to solve, should consolidate these ideas.

$$p(0) = 0.632 \cos(-18°.4) + 0.4$$

$$= 0.632 \cos(18°.4) + 0.4$$

$$= 0.6 + 0.4 = 1.0, \text{ as given}$$

If you are starting a new work period at this point, RE-READ the SUMMARY on the last page.

MASS-SPRING-DAMPER, AGAIN!

13.1

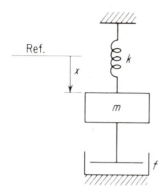

Ref.

x

k

m

f

As in section 7, the D.E. governing this system is

$$m \frac{d^2x}{dt^2} + f \frac{dx}{dt} + kx = mg,$$

with I.C.'s of $\left. \frac{dx}{dt} \right|_0$ and $x(0)$.

Therefore the transformed equation is:

_____

(*Don't manipulate the algebra.*)

13.2 SUBSTITUTE the data,

$$m = 2, \quad f = 4, \quad k = 8, \quad g = 8, \quad \left. \frac{dx}{dt} \right|_0 = 8, \quad \text{and} \quad x(0) = 0$$

and SOLVE for $X(s)$, without factoring the quadratic:

13.3 Now, FACTOR the quadratic $s^2 + 2s + 4$:

Answer: $s^2 + 2s + 4 = (s \qquad)(s \qquad)$

(*Check your answer by multiplying the factors together.*)

$$m\left\{s^2X(s) - sx(0) - \frac{dx}{dt}\Big|_0\right\} + f\{sX(s) - x(0)\} + kX(s) = mg/s$$

$$2\{s^2X(s) - 0 - 8\} + 4\{sX(s) - 0\} + 8X(s) = 16/s$$

$$2\{s^2 + 2s + 4\}X(s) = \frac{16}{s} + 16 = \frac{16(s + 1)}{s}$$

$$\therefore X(s) = \frac{8(s + 1)}{s(s^2 + 2s + 4)} \qquad \text{(or equivalent)}$$

$$(s + 1 - j\sqrt{3})(s + 1 + j\sqrt{3})$$

13.4   We have

$$X(s) = \frac{8(s + 1)}{s(s^2 + 2s + 4)} = \frac{8(s + 1)}{s(s + 1 - j\sqrt{3})(s + 1 + j\sqrt{3})}$$

$$= \frac{A}{s} + \frac{B}{s + 1 - j\sqrt{3}} + \frac{C}{s + 1 + j\sqrt{3}}$$

The next step is therefore to **FIND** A:

13.5   Then, **FIND** B in rectangular form:

13.6   **CONVERT** B into polar form:

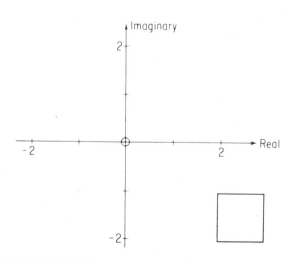

$$A = \left. \frac{8(s + 1)\cancel{s}}{\cancel{s}(s^2 + 2s + 4)} \right|_{s=0} = \underline{2}$$

$$B = \left. \frac{8(s + 1)\cancel{(s + 1 - j\sqrt{3})}}{s\cancel{(s + 1 - j\sqrt{3})}(s + 1 + j\sqrt{3})} \right|_{s = -1 + j\sqrt{3}}$$

$$= \frac{8(-1 + j\sqrt{3} + 1)}{(-1 + j\sqrt{3})(-1 + j\sqrt{3} + 1 + j\sqrt{3})} = \frac{8(j\sqrt{3})}{(-1 + j\sqrt{3})(2j\sqrt{3})} = \frac{4}{(-1 + j\sqrt{3})}$$

$$= \frac{4(-1 - j\sqrt{3})}{(-1 + j\sqrt{3})(-1 - j\sqrt{3})} = \frac{4(-1 - j\sqrt{3})}{4}$$

$$= \underline{-1 - j\sqrt{3}} \quad \text{or} \quad \underline{-1 - j1.73}$$

$$M = \sqrt{1^2 + (\sqrt{3})^2} = 2$$

and $\quad \theta = -90° - \tan^{-1} \dfrac{1}{\sqrt{3}}$

$$= -(90° + 30°)$$

$$= -120°$$

therefore $\quad B = \underline{2\epsilon^{-j120°}}$

(The answer $B = 2\epsilon^{j240°}$ is correct but less desirable. We usually keep $|\theta| \le 180°$.)

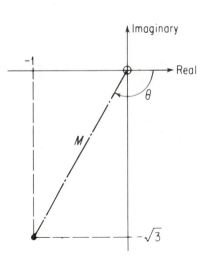

13.7   Having found $B = 2\epsilon^{-j120°}$, it follows that

$C =$ _____ (in polar form)

13.8   SUMMARIZE the results of the Partial Fraction expansion, making certain that the terms appear in the "correct" order:

$$X(s) = \frac{8(s + 1)}{s(s + 1 - j\sqrt{3})(s + 1 + j\sqrt{3})} = \underline{\hspace{6cm}}$$

13.9   Therefore

$x(t) =$ _____

CHECK that $x(0) = 0$, as given.

The next section (which is mostly review) will take you 20 to 40 minutes to complete.

$$C = \underline{2\epsilon^{+j120°}} \qquad \text{(complex conjugate of } B\text{)}$$

$$X(s) = \underline{\frac{2}{s} + \frac{2\epsilon^{-j120°}}{s + 1 - j\sqrt{3}} + \frac{2\epsilon^{+j120°}}{s + 1 + j\sqrt{3}}}$$

$$x(t) = \underline{2 + 4\epsilon^{-t}\cos(\sqrt{3}\,t - 120°)}$$

Check: $x(0) = 2 + 4\cos(-120°) = 2 - 2 = 0$

A SECTION OF REVIEW

We have been busy solving D.E.'s with the help of items on our semi-notes. It is time that we reviewed the background of some of these items.

14.1   *Without* looking at your semi-notes, DEFINE the L.T. of $f(t)$:

$$\mathcal{L}[f(t)] = F(s) =$$

_____   ☐

14.2   We can observe from the limits on the defining integral that only the behavior of $f(t)$ in the range of t between _____ and _____ can affect the evaluation of $F(s)$.

☐

14.3   Thus the statement that "the inverse transform of A/s^2 is At"
is valid *only* if t is _____ .

☐

14.4   In other words the statement that "the inverse transform of A/s^2 is At"
is *invalid* if t is _____ .

☐

14.5   All the problem solutions we have obtained are therefore valid only for t _____ .

☐

$$\mathcal{L}[f(t)] = F(s) = \int_0^\infty f(t)\,\epsilon^{-st}\,dt$$

$\underline{0}$ and $\underline{\infty}$

Comment:
We say that $\mathcal{L}[f(t)]$ depends only on the behavior of $f(t)$ for $t \geq 0$. The behavior of $f(t)$ for $t < 0$ does *not* affect $\mathcal{L}[f(t)]$.

$\underline{\text{zero or greater}}$ or $\underline{\geq 0}$ (*greater than zero*, or >0 is acceptable

but $t = 0$ is normally included.)

$\underline{\text{negative}}$ or $\underline{\text{less than zero}}$ or $\underline{<0}$

$\underline{\text{zero or greater}}$ or $\underline{\geq 0}$

121

14.6   COPY the statement "$t \geq 0$" at the end of line 1 of your semi-notes. NOTE that this restriction has also been indicated on line 4.

Comment:
As a matter of convention we will include t *equal* to zero in the range of validity.

14.7   We have used the notation $\mathcal{L}[\;\;]$ to represent the taking of the L.T. It is conventional to extend the notation to the taking of the inverse transform by writing $\mathcal{L}^{-1}[\;\;]$.

COPY the statement

$$\mathcal{L}^{-1}[F(s)] = f(t) \quad t \geq 0$$

onto line 3 of your semi-notes.

14.8 $\mathcal{L}^{-1}[10/(s + 1)] = \underline{\hspace{3cm}}$, valid for t $\underline{\hspace{3cm}}$.

(*You may use your semi-notes.*)

14.9   For further review, use your semi-notes to EVALUATE

$$\mathcal{L}[7t^4] = \underline{\hspace{4cm}}$$

14.10   Similarly, from your semi-notes,

$$\mathcal{L}^{-1}[12/s^4] = \underline{\hspace{3cm}} \quad \text{valid for} \quad \underline{\hspace{3cm}}$$

$\mathcal{L}^{-1}[10/(s + 1)] = \underline{10\epsilon^{-t}}$ valid for $t \geq 0$

This is usually written

$\mathcal{L}^{-1}[10/(s + 1)] = 10\epsilon^{-t}$ $t \geq 0$

$\mathcal{L}[7t^4] = \dfrac{7(4!)}{s^5}$ or $\dfrac{7(24)}{s^5}$ or $\dfrac{168}{s^5}$

$\mathcal{L}^{-1}[12/s^4] = \mathcal{L}^{-1}[2(3!)/s^4] = \underline{2t^3}$ $\underline{t \geq 0}$

14.11   Use the facts that $\mathcal{L}[A\epsilon^{at}] = A/(s - a)$ and that $\sinh \alpha t = \frac{1}{2}(\epsilon^{\alpha t} - \epsilon^{-\alpha t})$ to FIND

$$\mathcal{L}[A \sinh \alpha t] =$$

$$=$$

$$=$$
_____

14.12   With the help of the result $\int_a^b u \, dv = uv \Big|_a^b - \int_a^b v \, du$, and without reference to your earlier derivation, COMPLETE the derivation of Theorem 2:

$$\mathcal{L}\left[\frac{df(t)}{dt}\right] = \int_0^\infty \epsilon^{-st} \frac{df(t)}{dt} \, dt$$

$$=$$

$$=$$

$$=$$

$$= sF(s) - f(0)$$

$$\mathcal{L}[A \sinh \alpha t] = \mathcal{L}\left[\frac{A}{2}(\epsilon^{\alpha t} - \epsilon^{-\alpha t})\right]$$

$$= \frac{A}{2}\left(\frac{1}{s - \alpha} - \frac{1}{s + \alpha}\right)$$

$$= \frac{A}{2}\frac{s + \alpha - s + \alpha}{s^2 - \alpha^2}$$

$$= \frac{A\alpha}{s^2 - \alpha^2} \qquad \text{or equivalent}$$

In $\displaystyle\int_0^\infty \epsilon^{-st}\frac{df(t)}{dt}\,dt$, $u = \epsilon^{-st}$ and $dv = \dfrac{df(t)}{dt}\,dt$

note!

Therefore $du = -s\epsilon^{-st}\,dt$ and $v = f(t)$.

Thus, substituting into the formula for integration by parts:

$$\int_0^\infty \epsilon^{-st}\frac{df(t)}{dt}\,dt = \epsilon^{-st}f(t)\Big|_0^\infty - \int_0^\infty f(t)\{-s\epsilon^{-st}\,dt\}$$

$$= \epsilon^{-s\infty}f(\infty) - \epsilon^{-s0}f(0) + s\int_0^\infty f(t)\epsilon^{-st}\,dt$$

$$= 0 - f(0) + sF(s)$$

or $\qquad \underline{\mathcal{L}[df/dt] = sF(s) - f(0)} \qquad$ (for functions such that $\epsilon^{-s\infty}f(\infty) \to 0$ for some finite s.)

14.13 DERIVE Theorem 3 (that is, FIND $\mathcal{L}[d^2f/dt^2]$) with the aid of Theorem 2 and the auxiliary function $g(t) = df/dt$:

Don't forget that we use the notation $df/dt\,|_0$ for $g(0)$.

☐

14.14 Go back and REVIEW the derivation of the result

$$\mathcal{L}\left[\int_0^t f(t)\,dt\right] = F(s)/s$$

in frames **8.1** through **8.3**.

The next section will take you 20 to 40 minutes to complete.

$$d^2f/dt^2 = dg/dt \qquad\qquad \text{where} \quad g(t) = df/dt$$

thus $\quad \mathcal{L}[d^2f/dt^2] = \mathcal{L}[dg/dt] \qquad$ and $\quad \mathcal{L}[g(t)] = \mathcal{L}[df/dt]$

$$= sG(s) - g(0) \qquad\qquad \text{or} \qquad G(s) = sF(s) - f(0)$$

Substituting for $G(s)$ and $g(0)$:

$$\mathcal{L}[d^2f/dt^2] = s\{sF(s) - f(0)\} - df/dt\,|_0$$

$$= s^2F(s) - sf(0) - df/dt\,|_0$$

HEATING AN OVEN

Here an electric heating coil puts heat, $q_i(t)$, into the oven, while $q_0(t)$ leaks out through the insulation. We will solve for the temperature difference between the inside and the outside of the oven, $\theta(t)$.

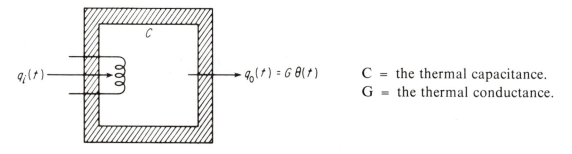

C = the thermal capacitance.
G = the thermal conductance.

The D.E. in this case is:

$$C \frac{d\theta(t)}{dt} + G\theta(t) = q_i(t) \quad \text{with the initial condition } \theta(0)$$

and its transform is

$$C\{s\Theta(s) - \theta(0)\} + G\Theta(s) = Q_i(s)$$

15.1 SUBSTITUTE the data:

$$C = 1, \quad G = 1, \quad \theta(0) = 2, \quad \text{and} \quad q_i(t) = 5t^2$$

and SOLVE for $\Theta(s)$:

$$Q_i(s) = (5)(2!)/s^3 \quad \text{or} \quad 10/s^3$$

$$(s + 1)\Theta(s) = 10/s^3 + 2$$

$$\Theta(s) = \frac{2(s^3 + 5)}{s^3(s + 1)}$$

15.2   From the previous frame we have $\Theta(s) = \dfrac{2(s^3 + 5)}{s^3(s + 1)}$ and $\Theta(s)$ is then said to have a "3-fold *repeated root*" in its denominator.

If we consider the process of combining terms over a common denominator, we see that the most general form of the *reverse* process—P.F. expansion—must be

$$\Theta(s) = \frac{2(s^3 + 5)}{s^3(s + 1)} = \frac{A}{s + 1} + \frac{B}{s^3} + \frac{C}{s^2} + \frac{D}{s}.$$

Then $A =$ $=$ $=$ _____

and $B =$ $=$ $=$ _____

$$A = \frac{2(s^3 + 5)(\cancel{s + 1})}{s^3(\cancel{s + 1})}\Bigg|_{s = -1} = \frac{2(4)}{-1} = \underline{-8}$$

$$B = \frac{2(s^3 + 5)\cancel{s^3}}{\cancel{s^3}(s + 1)}\Bigg|_{s = 0} = \frac{2(5)}{1} = \underline{10}$$

15.3   If we now try to find C in the normal way, we first multiply by s^2:

$$\frac{2(s^3 + 5)\cancel{s}^2}{\cancel{s^3}(s + 1)} = \frac{As^2}{s + 1} + \frac{B\cancel{s}^2}{\cancel{s^3}} + \frac{C\cancel{s}^2}{\cancel{s^2}} + \frac{Ds^{\cancel{2}}}{\cancel{s}}$$

Then, if we put $s = 0$, as usual, we find that the term on the left and the second term on the right go to infinity! The first and last terms on the right go to zero, in the normal fashion.

However, if we simply combine the two misbehaving terms, knowing that $B = 10$, we find that:

$$C = \left\{ \frac{2(s^3 + 5)\cancel{s}^2}{\cancel{s^3}(s + 1)} - \frac{B\cancel{s}^2}{\cancel{s^3}} \right\}\Bigg|_{s=0} = \left\{ \frac{2(s^3 + 5)}{s(s + 1)} - \frac{10}{s} \right\}\Bigg|_{s=0}$$

$$= \frac{\underline{\hspace{4cm}}}{s(s + 1)}\Bigg|_{s=0}$$
                               (combine the two terms
over the common denominator
and cancel the offending s.)

$$= \frac{\underline{\hspace{3cm}}}{\cancel{s}(s + 1)}\Bigg|_{s=0} = \underline{\hspace{3cm}}$$

15.4   With the value of C as above, and with $B = 10$, FIND D, using the method above, in the expansion

$$\frac{2(s^3 + 5)}{s^3(s + 1)} = \frac{A}{s + 1} + \frac{B}{s^3} + \frac{C}{s^2} + \frac{D}{s}.$$

$$C = \frac{2s^3 + 10 - 10s - 10}{s(s+1)}\Bigg|_{s=0}$$

$$= \frac{2s(s^2 - 5)}{s(s+1)}\Bigg|_{s=0} = \frac{2(-5)}{1} \qquad \text{(note the cancellation of } s\text{)}$$

$$C = \underline{-10}$$

Comment:
Because s cancels in the combination of the two terms which individually become infinite as $s \longrightarrow 0$, the *combination* has a finite value (-10).

We find D (as usual) by multiplying by the denominator of the term in D, in this case s, and by then putting $s = 0$ (in this example):

$$D = \left\{ \frac{2(s^3 + 5)}{s^2(s+1)} - \frac{B}{s^2} - \frac{C}{s} \right\}\Bigg|_{s=0}$$

$$= \frac{2s^3 + 10 - 10s - 10 + 10s^2 + 10s}{s^2(s+1)}\Bigg|_{s=0}$$

$$= \frac{2s^2(s + 5)}{s^2(s+1)}\Bigg|_{s=0} = \frac{2(5)}{1} = \underline{10}$$

A COMMENT

15.5   The repeated denominator root in the P.F. expansion

$$\frac{2(s^3 + 5)}{s^3(s + 1)} = \frac{A}{s + 1} + \frac{B}{s^3} + \frac{C}{s^2} + \frac{D}{s}$$

necessitated the exercise of additional care, but the expansion was still handled by the normal technique.

The only "trick" you should note is that the coefficients corresponding to the repeated root (B, C, and D, above) should be calculated *in order*, starting with the coefficient of the term with the *highest* denominator power.

15.6   Combining the results from the last frames:

$$\Theta(s) = \frac{2(s^3 + 5)}{s^3(s + 1)} = \frac{-8}{s + 1} + \frac{10}{s^3} - \frac{10}{s^2} + \frac{10}{s}.$$

Therefore,

$$\theta(t) = \mathcal{L}^{-1}[\Theta(s)] = \underline{\hspace{5cm}}$$

which is valid for t $\underline{\hspace{3cm}}$.

CHECK your result against the I.C., $\theta(0) = 2$:

$$\theta(t) = \mathcal{L}^{-1}[\Theta(s)] = \underline{-8\epsilon^{-t} + 5t^2 - 10t + 10}$$

which is valid for $t \geq 0$.

Check: $\theta(0) = -8 + 10 = 2$

ANOTHER EXAMPLE OF REPEATED ROOTS

15.7 For additional practice in the handling of repeated roots, we will find the inverse transform of

$$Y(s) = \frac{1}{s(s+1)^2} = \frac{A}{s} + \frac{B}{(s+1)^2} + \frac{C}{s+1}.$$

First, EVALUATE A and B:

15.8 Next, COMPUTE the value of C:

15.9 Given that $\mathcal{L}^{-1}[1/(s+1)^2] = t\epsilon^{-t}$, and your calculated values of A, B, and C in the above expansion, you can write down

$$y(t) = \mathcal{L}^{-1}[Y(s)] = \underline{\hspace{4cm}} \text{ valid for } \underline{\hspace{1.5cm}}$$

$$A = \frac{1(\cancel{s})}{\cancel{s}(s+1)^2}\Bigg|_{s=0} = \frac{1}{1} = \underline{1}$$

$$B = \frac{1\cancel{(s+1)^2}}{s\cancel{(s+1)^2}}\Bigg|_{s=-1} = \frac{1}{-1} = \underline{-1}$$

Comment:
We *are* calculating the coefficients in the "correct" order—the coefficient B, corresponding to the *highest* order of the repeated root, before C.

We have to be more careful when finding C. Thus:

$$\frac{1\cancel{(s+1)}}{s(s+1)^{\cancel{2}}} = \frac{A(s+1)}{s} + \frac{B\cancel{(s+1)}}{(s+1)^{\cancel{2}}} + \frac{C\cancel{(s+1)}}{\cancel{(s+1)}}$$

Only the first term on the right hand side goes to zero when we put $s = -1$, leaving:

$$C = \left\{\frac{1}{s(s+1)} - \frac{B}{s+1}\right\}\Bigg|_{s=-1} = \left\{\frac{1}{s(s+1)} - \frac{-1}{s+1}\right\}\Bigg|_{s=-1}$$

$$= \frac{\cancel{(1+s)}}{s\cancel{(s+1)}}\Bigg|_{s=-1} \qquad \text{(combining the two terms over}$$
$$\text{the common denominator)}$$

$$= \frac{1}{-1} = \underline{-1}$$

$$\mathcal{L}^{-1}[1/s(s+1)^2] = \mathcal{L}^{-1}[1/s - 1/(s+1)^2 - 1/(s+1)]$$

$$= \underline{1 - t\epsilon^{-t} - \epsilon^{-t}} \qquad \text{valid for } \underline{t \geq 0}$$

This section will take you 10–25 minutes.

THE LAST THEOREM

16.1   We have just solved a problem with a repeated root, $(s + 1)^2$. In the process we needed the result $\mathcal{L}^{-1}[1/(s + 1)^2] = t\epsilon^{-t}$.

Here we will consider a more general question: The evaluation of $\mathcal{L}[\epsilon^{at} f(t)]$. This clearly covers the calculation of $\mathcal{L}[\epsilon^{-t}t]$ as a special case.

By definition,

$$\mathcal{L}[\epsilon^{at} f(t)] = \int_0^\infty \{\epsilon^{at} f(t)\} \epsilon^{-st} \, dt$$

$$= \int_0^\infty f(t) \, \epsilon^{-(s-a)t} \, dt$$

$$= \underline{F(\qquad\qquad)}$$

Hint: Compare the integral above with $F(s) = \displaystyle\int_0^\infty f(t) \epsilon^{-st} \, dt$

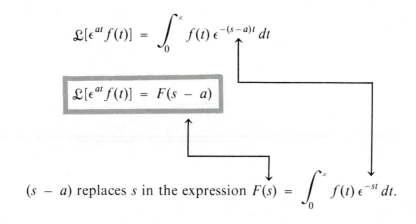

$$\mathcal{L}[\epsilon^{at} f(t)] = \int_0^\infty f(t)\, \epsilon^{-(s-a)t}\, dt$$

$$\boxed{\mathcal{L}[\epsilon^{at} f(t)] = F(s - a)}$$

$(s - a)$ replaces s in the expression $F(s) = \displaystyle\int_0^\infty f(t)\, \epsilon^{-st}\, dt.$

16.2 COPY the result

$$\mathcal{L}[\epsilon^{at} f(t)] = F(s - a)$$

onto line 17 of your semi-notes, as Theorem 6.

16.3 Theorem 6 sometimes appears puzzlingly simple: It *is* simple, and you must not let it frighten you!

Suppose we wish to evaluate $\mathcal{L}[\epsilon^{at} t]$, using Theorem 6:
Here $f(t)$ is t, and we know that

$$\mathcal{L}[t] = 1/s^2$$

$$\therefore \quad \mathcal{L}[\epsilon^{at} t] = \underline{\hspace{4cm}}$$

16.4 Specifically,

$$\mathcal{L}[\epsilon^{-t} t] = \underline{\hspace{4cm}}$$

16.5 Similarly, given that $\mathcal{L}[t^2] = 2!/s^3$

then $\quad \mathcal{L}[\epsilon^{3t} t^2] = \underline{\hspace{4cm}}$

Knowing that $\mathcal{L}[t] = 1/s^2$

$$\mathcal{L}[\epsilon^{at} t] = \underline{1/(s - a)^2}$$

The theorem simply asks you to replace s by $s - a$

Here $a = -1$ in the result immediately above,

therefore $\mathcal{L}[\epsilon^{-t} t] = \underline{1/(s + 1)^2}$

which was the result we used at the end of the last section.

$$\mathcal{L}[\epsilon^{3t} t^2] = \underline{2!/(s - 3)^3} \quad \text{or} \quad \underline{2/(s - 3)^3} \quad \text{(replace } s \text{ by } s - 3)$$

16.6   Knowing that $\mathcal{L}[5 \sin 7t] = \dfrac{(5)(7)}{s^2 + 7^2} = \dfrac{35}{s^2 + 7^2}$

it follows from Theorem 6 that

$\mathcal{L}[5\epsilon^{9t} \sin 7t] =$

16.7   Similarly,

$\mathcal{L}[2\epsilon^{3t} \cos 4t] =$

(*Be careful:* substitute for s everywhere!)

16.8   We found earlier, in frame **10.11**, that

$$\mathcal{L}[A \cos (\omega t + \theta)] = \frac{A \cos \theta (s - \omega \tan \theta)}{s^2 + \omega^2}$$

where A, ω, θ, $\cos \theta$, and $\tan \theta$ are constants.

Therefore, $\mathcal{L}[A\epsilon^{\alpha t} \cos (\omega t + \theta)] =$

If $\quad \mathcal{L}[5 \sin 7t] = \dfrac{35}{s^2 + 7^2}$

then $\quad \mathcal{L}[5\epsilon^{9t} \sin 7t] = \dfrac{35}{(s - 9)^2 + 7^2} \qquad$ or equivalent

---

$$\mathcal{L}[2 \cos 4t] = \dfrac{2s}{s^2 + 4^2}$$

$$\therefore \mathcal{L}[2\epsilon^{3t} \cos 4t] = \dfrac{2(s - 3)}{(s - 3)^2 + 4^2} \qquad \text{or equivalent}$$

$$\mathcal{L}[A\epsilon^{\alpha t} \cos (\omega t + \theta)] = \dfrac{A \cos \theta (s - \alpha - \omega \tan \theta)}{(s - \alpha)^2 + \omega^2} \qquad \text{or equivalent}$$

143

16.9   If we replace A by $2M$ in the last result we obtain the transform pair

$$2M\epsilon^{\alpha t} \cos(\omega t + \theta) \longleftrightarrow \frac{2M\cos\theta(s - \alpha - \omega\tan\theta)}{(s - \alpha)^2 + \omega^2}$$

WRITE "$= \dfrac{2M\cos\theta(s - \alpha - \omega\tan\theta)}{(s - \alpha)^2 + \omega^2}$"

at the end of line 11 of your semi-notes, then READ the comment opposite.

16.10   We most commonly apply Theorem 6 when evaluating *inverse* transforms.   It is then more convenient to "reverse" Theorem 6.   That is,

$$\mathcal{L}^{-1}[F(s - a)] = \underline{\hspace{4cm}} \qquad t \geq 0$$

16.11   COPY the above result onto line 18 of your semi-notes as Theorem 6a.
Then EVALUATE:

$$\mathcal{L}^{-1}[5/(s + 6)^2] = \underline{\hspace{4cm}} \qquad t \geq 0$$

16.12   Now, try this one, carefully!   (Use your semi-notes.)

$$\mathcal{L}^{-1}[An!/(s - \alpha)^{n+1}] = \underline{\hspace{3cm}} \qquad t \geq 0$$

16.13   And finally, $\mathcal{L}^{-1}\left[\dfrac{6(s + 17)}{(s + 17)^2 + 4^2}\right] = \underline{\hspace{4cm}} \qquad t \geq 0$

144

Comment:
We are saying that if we were to combine the two terms in the righthand column of the table, we would obtain the new result. Try this if you like!

Since $\quad F(s - a) = \mathcal{L}[\epsilon^{at} f(t)]$

$$\boxed{\mathcal{L}^{-1}[F(s - a)] = \epsilon^{at} f(t) \qquad t \geq 0}$$

In this example $a = -6$ and $F(s)$ must be $5/s^2$. Therefore $f(t) = 5t$

and $\quad \mathcal{L}^{-1}[5/(s + 6)^2] = \underline{5\epsilon^{-6t} t} \qquad t \geq 0$

$$\mathcal{L}^{-1}[An!/(s - \alpha)^{n+1}] = \underline{A\epsilon^{\alpha t} t^n} \qquad t \geq 0$$

$$\mathcal{L}^{-1}\left[\frac{6(s + 17)}{(s + 17)^2 + 4^2}\right] = \underline{6\epsilon^{-17t} \cos 4t} \qquad t \geq 0$$

In this example you will encounter repeated roots, and will therefore have another opportunity to apply Theorem 6a. The time for the section is 15 to 30 minutes.

PARALLEL L-R-C CIRCUIT WITH REPEATED ROOTS

17.1   In the following problem we will solve for $v(t)$, $t \geq 0$:

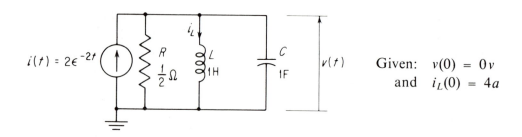

Given: $v(0) = 0v$

and $i_L(0) = 4a$

The governing equation is: $C\dfrac{dv}{dt} + \dfrac{v}{R} + \dfrac{1}{L}\displaystyle\int_0^t v\,dt + i_L(0) = i(t),$

which transforms to: $C\{sV(s) - v(0)\} + \dfrac{V(s)}{R} + \dfrac{V(s)}{Ls} + \dfrac{i_L(0)}{s} = I(s).$

Substituting the given data* into the transformed equation:

$$\{s + 2 + 1/s\}V(s) = 2/(s + 2) - 4/s = \dfrac{2s - 4s - 8}{s(s + 2)}$$

or $$V(s) = \dfrac{-2(s + 4)}{(s^2 + 2s + 1)(s + 2)}$$

$$= \dfrac{-2(s + 4)}{(\qquad)^2(s + 2)} = \dfrac{A}{\qquad} + \dfrac{B}{\qquad} + \dfrac{C}{s + 2}$$

(COMPLETE the last line, above.)

---

*We will find repeated roots only because of a special choice of parameter values. In practice the roots will never be *exactly* equal, due to small variations in the actual parameter values.

$$V(s) = \frac{-2(s + 4)}{(s^2 + 2s + 1)(s + 2)}$$

$$= \frac{-2(s + 4)}{(s + 1)^2(s + 2)} = \frac{A}{(s + 1)^2} + \frac{B}{s + 1} + \frac{C}{s + 2} \qquad \text{or equivalent}$$

17.2 RE-READ the comment on frame **15.5** and then FIND A, B, and C in the expansion

$$V(s) = \frac{-2(s + 4)}{(s + 1)^2(s + 2)} = \frac{A}{(s + 1)^2} + \frac{B}{s + 1} + \frac{C}{s + 2}.$$

$$A = \left.\frac{-2(s+4)(s+1)^2}{(s+1)^2(s+2)}\right|_{s=-1} = \frac{-2(3)}{1} = \underline{-6} \qquad \text{(we must find } A \text{ before solving for } B\text{)}$$

$$B = \left.\left\{\frac{-2(s+4)(s+1)}{(s+1)^2(s+2)} - \frac{A(s+1)}{(s+1)^2}\right\}\right|_{s=-1} \qquad \text{(where } A \text{ is now known)}$$

$$= \left.\frac{-2s - 8 + 6s + 12}{(s+1)(s+2)}\right|_{s=-1} = \left.\frac{4(s+1)}{(s+1)(s+2)}\right|_{s=-1} = \underline{4}$$

$$\text{and} \quad C = \left.\frac{-2(s+4)(s+2)}{(s+1)^2(s+2)}\right|_{s=-2} = \frac{-2(2)}{(-1)^2} = \underline{-4}$$

Comment:
The expansion is easily checked by recombining the terms over a common denominator.

17.3   We now have $V(s) = \dfrac{-6}{(s + 1)^2} + \dfrac{4}{s + 1} - \dfrac{4}{s + 2}$.

Therefore $v(t) =$ _____   volt $t \geq 0$.

CHECK your result against the fact that $v(0) = 0\,v$.

SOME MORE REPEATED ROOTS!

17.4   As another example of repeated roots, FIND A, B, and C in the expansion

$$Q(s) = \frac{s^3 + 2s^2 - 2}{s^2(s + 2)^2} = \frac{A}{s^2} + \frac{B}{s} + \frac{C}{(s + 2)^2} + \frac{D}{s + 2}.$$

(*Don't* try to find D in this frame.)

$$\mathcal{L}[At] = A/s^2 \text{ and Theorem 6a.}$$

$$v(t) = \underline{-6t\epsilon^{-t} + 4\epsilon^{-t} - 4\epsilon^{-2t}} \text{ volt} \qquad t \geq 0$$

Check: $\qquad v(0) = 0 + 4 - 4 = 0v.$

$$\frac{s^3 + 2s^2 - 2}{s^2(s + 2)^2} = \frac{A}{s^2} + \frac{B}{s} + \frac{C}{(s + 2)^2} + \frac{D}{s + 2}$$

Highest denominator powers first

$$A = \left.\frac{(s^3 + 2s^2 - 2)s^2}{s^2(s + 2)^2}\right|_{s=0} = \frac{-2}{4} = -\frac{1}{2}$$

$$B = \left.\left\{\frac{(s^3 + 2s^2 - 2)s}{s^2(s + 2)^2} - \frac{As}{s^2}\right\}\right|_{s=0} = \left.\left\{\frac{s^3 + 2s^2 - 2}{s(s + 2)^2} + \frac{\frac{1}{2}}{s}\right\}\right|_{s=0}$$

$$= \left.\frac{s^3 + 2s^2 - 2 + \frac{1}{2}s^2 + 2s + 2}{s(s + 2)^2}\right|_{s=0} = \left.\frac{s(s^2 + 2.5s + 2)}{s(s + 2)^2}\right|_{s=0} = \frac{2}{4} = \frac{1}{2}$$

$$C = \left.\frac{(s^3 + 2s^2 - 2)(s + 2)^2}{s^2(s + 2)^2}\right|_{s=-2} = \frac{-2}{4} = -\frac{1}{2}$$

$$A = -\frac{1}{2}; \quad B = \frac{1}{2}; \quad \text{and} \quad C = -\frac{1}{2}$$

17.5   We now have

$$Q(s) = \frac{s^3 + 2s^2 - 2}{s^2(s+2)^2} = \frac{-\frac{1}{2}}{s^2} + \frac{\frac{1}{2}}{s} - \frac{\frac{1}{2}}{(s+2)^2} + \frac{D}{s+2}$$

Proceeding as usual,

$$D = \left\{ \frac{(s^3 + 2s^2 - 2)(s+2)}{s^2(s+2)^2} - \frac{-\frac{1}{2}(s+2)}{(s+2)^2} \right\} \Bigg|_{s=-2}$$

$$= \frac{s^3 + 2s^2 - 2 + \frac{1}{2}s^2}{s^2(s+2)} \Bigg|_{s=-2} = \frac{s^3 + 2.5s^2 - 2}{s^2(s+2)} \Bigg|_{s=-2}$$

We realize that $s + 2$ must cancel, but the result of dividing $(s + 2)$ into the numerator is not obvious. You must therefore COMPLETE the following long division before finally determining D:

$$
\begin{array}{r}
s^2 \\
s + 2 \overline{\smash{\big)}\ s^3 + 2.5s^2 + 0s - 2} \\
\underline{s^3 + 2.0s^2 } \\
+ 0.5s^2 + 0s \\
\underline{} \\
\end{array}
$$

If all our work is correct, you should have zero remainder after completing the division.

Now, COMPLETE the evaluation of D:

$$D = \frac{s^3 + 2.5s^2 - 2}{s^2(s+2)} \Bigg|_{s=-2} = \frac{(s+2)()}{s^2(s+2)} \Bigg|_{s=-2}$$

$$= \underline{}$$

$$
\begin{array}{r}
s^2 + 0.5s \; - 1 \\
s + 2 \; \big)\overline{\; s^3 + 2.5s^2 + 0s - 2} \\
\underline{s^3 + 2.0s^2} \\
0.5s^2 + 0s \\
\underline{0.5s^2 + \; s} \\
-s - 2 \\
\underline{-s - 2} \\
0 \quad 0
\end{array}
$$

$$
D = \left. \frac{(s + 2)(s^2 + 0.5s - 1)}{s^2(s + 2)} \right|_{s = -2} = \frac{(4 - 1 - 1)}{4} = \underline{\frac{1}{2}}
$$

Comment:

We will not bother with the final (easy) step, that of writing down $q(t) = \mathcal{L}^{-1}[Q(s)]$.

This section, where you will solve a pair of simultaneous D.E.'s, will take you 20 to 40 minutes to work.

AUTOMOBILE FLUID-COUPLED TRANSMISSION

18.1   Here we will solve for the crankshaft speed, ω_e, and the drive-shaft speed, ω_d, for $t \geq 0$. We are given that $\omega_e(0) = \omega_d(0) = 0$.

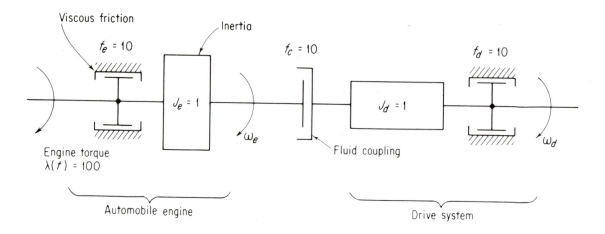

Equating the torques on each of the two shafts to zero, we obtain *two* equations in $\omega_e(t)$ and $\omega_d(t)$. If we transform them, substitute the data, and rearrange we find that

$$(s + 20)\Omega_e(s) - 10\Omega_d(s) = 100/s \qquad \text{where} \quad \mathcal{L}[\omega_e(t)] = \Omega_e(s)$$

$$-10\Omega_e(s) + (s + 20)\Omega_d(s) = 0 \qquad \text{and} \quad \mathcal{L}[\omega_d(t)] = \Omega_d(s)$$

READ THE COMMENTS OPPOSITE

Comments:

1. Ω and ω are capital and lower case "omega."
2. If you prefer matrix notation, the equations are:

$$\begin{bmatrix} s + 20 & -10 \\ -10 & s + 20 \end{bmatrix} \begin{bmatrix} \Omega_e(s) \\ \Omega_d(s) \end{bmatrix} = \begin{bmatrix} 100/s \\ 0 \end{bmatrix}$$

18.2 We have that

$$(s + 20)\Omega_e(s) - 10\Omega_d(s) = 100/s$$
$$-10\Omega_e(s) + (s + 20)\Omega_d(s) = 0$$

The next step is the standard one of solving for the algebraic unknown(s). Hitherto there has been only one unknown in each problem—here we have two. Obviously we must now solve the *simultaneous* algebraic equations above for the two unknowns, $\Omega_e(s)$ and $\Omega_d(s)$.

Thus, solving by determinants (Cramer's rule):

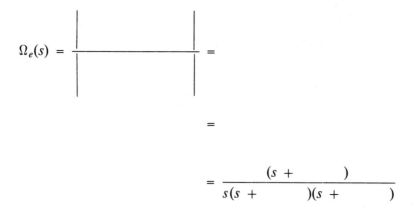

$$\Omega_e(s) = \frac{\begin{vmatrix} & \\ & \end{vmatrix}}{\begin{vmatrix} & \\ & \end{vmatrix}} =$$

$$=$$

$$= \frac{(s + \quad)}{s(s + \quad)(s + \quad)}$$

$\square$

18.3 Similarly,

$$\Omega_d(s) = \frac{\begin{vmatrix} & \\ & \end{vmatrix}}{\begin{vmatrix} & \\ & \end{vmatrix}} =$$

$$=$$

$$= \frac{}{s(s + \quad)(s + \quad)}$$

$\square$

$$\Omega_e(s) = \frac{\begin{vmatrix} 100/s & -10 \\ 0 & s+20 \end{vmatrix}}{\begin{vmatrix} s+20 & -10 \\ -10 & s+20 \end{vmatrix}} = \frac{100(s+20)/s}{(s+20)(s+20) - 100}$$

$$= \frac{100(s+20)}{s(s^2 + 40s + 300)}$$

or
$$\Omega_e(s) = \frac{100(s+20)}{s(s+10)(s+30)}$$

$$\Omega_d(s) = \frac{\begin{vmatrix} s+20 & 100/s \\ -10 & 0 \end{vmatrix}}{\begin{vmatrix} s+20 & -10 \\ -10 & s+20 \end{vmatrix}} = \frac{1000/s}{(s+20)(s+20) - 100}$$

$$= \frac{1000}{s(s^2 + 40s + 300)}$$

or
$$\Omega_d(s) = \frac{1000}{s(s+10)(s+30)}$$

18.4 Knowing that $\Omega_e(s) = \dfrac{100(s + 20)}{s(s + 10)(s + 30)}$, FIND $\omega_e(t)$ for $t \geq 0$:

Expanding in P.F.'s: $\Omega_e(s) = \dfrac{100(s + 20)}{s(s + 10)(s + 30)} = \dfrac{A}{s} + \dfrac{B}{s + 10} + \dfrac{C}{s + 30}$

where $\quad A = \dfrac{100(s + 20)\not{s}}{\not{s}(s + 10)(s + 30)}\bigg|_{s=0} = \dfrac{100(20)}{10(30)} = 6.67$

$\qquad\quad B = \dfrac{100(s + 20)(\not{s + 10})}{s(\not{s + 10})(s + 30)}\bigg|_{s=-10} = \dfrac{100(10)}{-10(20)} = -5$

and $\qquad C = \dfrac{100(s + 20)(\not{s + 30})}{s(s + 10)(\not{s + 30})}\bigg|_{s=-30} = \dfrac{100(-10)}{-30(-20)} = -1.67$

Therefore $\underline{\omega_e(t) = 6.67 - 5\epsilon^{-10t} - 1.67\epsilon^{-30t}} \qquad t \geq 0$

Check: $\quad \omega_e(0) = 6.67 - 5 - 1.67 = 0$

18.5 Finally, having $\Omega_d(s) = \dfrac{1000}{s(s + 10)(s + 30)}$, FIND $\omega_d(t)$ for $t \geq 0$:

SUMMARY

We have seen that the method of solving *simultaneous* D.E.'s is entirely obvious. That is, we

1. Transform *each* of the differential equations;
2. Solve the resulting *simultaneous* algebraic equations for the algebraic unknowns; and
3. Take the inverse transform of *each* of the algebraic unknowns in the usual way.

Expanding in P.F.'s: $\Omega_d(s) = \dfrac{1000}{s(s + 10)(s + 30)} = \dfrac{A}{s} + \dfrac{B}{s + 10} + \dfrac{C}{s + 30}$

where $A = \dfrac{1000\not{s}}{\not{s}(s + 10)(s + 30)}\bigg|_{s=0} = \dfrac{1000}{10(30)} = 3.33$

$B = \dfrac{1000(s + 10)}{s(s + 10)(s + 30)}\bigg|_{s=-10} = \dfrac{1000}{-10(20)} = -5$

and $C = \dfrac{1000(s + 30)}{s(s + 10)(s + 30)}\bigg|_{s=-30} = \dfrac{1000}{-30(-20)} = 1.67$

Thus $\omega_d(t) = 3.33 - 5\epsilon^{-10t} + 1.67\epsilon^{-30t}$ $t \geq 0$

Check: $\omega_d(0) = 3.33 - 5 + 1.67 = 0$

The working time for this section will be from 25 to 30 minutes.

A FINAL REVIEW

19.1   Without looking back at earlier sections, DERIVE Theorem 5.   That is, FIND $\mathcal{L}\left[\int_0^t f(t)\, dt\right]$ with the aid of Theorem 2 (see semi-notes) and the auxiliary function defined by

$$f(t) = dg/dt \qquad\qquad (1)$$

First, $\displaystyle\int_0^t f(t)\, dt = \int_0^t \frac{dg(t)}{dt}\, dt$

$$=$$

$$= \qquad\qquad (2)$$

Then, taking the L.T.:

$$f(t) = dg/dt \tag{1}$$

First, $\displaystyle\int_0^t f(t)\,dt = \int_0^t \frac{dg(t)}{dt}\,dt$

$$= g(t)\Big|_0^t$$

$$\therefore \int_0^t f(t)\,dt = g(t) - g(0) \tag{2}$$

Then, taking the L.T. of (2)

$$\mathcal{L}\left[\int_0^t f(t)\,dt\right] = G(s) - g(0)/s \tag{3}$$

Finally, taking the L.T. of (1)

$$F(s) = sG(s) - g(0) \tag{4}$$

Comparing (3) and (4) we see that

$$\mathcal{L}\left[\int_0^t f(t)\,dt\right] = F(s)/s$$

19.2 *Use Theorem 6* and your table of transform pairs to EVALUATE

$$\mathcal{L}[2\,t\epsilon^{3t}] = \underline{\hspace{3cm}}$$

☐

19.3 Then, *using the above result* and *Theorem 5*, EVALUATE

$$\mathcal{L}\left[\int_0^t 2t\epsilon^{3t}\,dt\right] = \underline{\hspace{3cm}}$$

☐

19.4 Also, EVALUATE $\mathcal{L}^{-1}\left[\dfrac{3}{s(s-2)^2}\right]$, *after expanding in P.F.'s:*

☐

We know that

$$\mathcal{L}[2t] = 2/s^2$$

Thus $\quad \mathcal{L}[2t\epsilon^{3t}] = \mathcal{L}[2\epsilon^{3t}t] = \underline{2/(s-3)^2}$

Given that

$$\mathcal{L}[2t\epsilon^{3t}] = 2/(s-3)^2$$

then $\quad \mathcal{L}\left[\int_0^t 2t\epsilon^{3t}\,dt\right] = \dfrac{2}{s(s-3)^2}$

$$\frac{3}{s(s-2)^2} = \frac{A}{s} + \frac{B}{(s-2)^2} + \frac{C}{s-2}$$

$$\therefore A = \left.\frac{3\not s}{\not s(s-2)^2}\right|_{s=0} = \frac{3}{4} = 0.75$$

$$B = \left.\frac{3(\overline{s-2})^2}{s(\overline{s-2})^2}\right|_{s=2} = \frac{3}{2} = 1.5$$

$$C = \left\{\frac{3(\overline{s-2})}{s(s-2)^2} - \frac{B(\overline{s-2})}{(s-2)^2}\right\}\Bigg|_{s=2} = \left.\frac{3-1.5s}{s(s-2)}\right|_{s=2}$$

$$= \left.\frac{-1.5(\overline{s-2})}{s(\overline{s-2})}\right|_{s=2} = \frac{-1.5}{2} = -0.75$$

$$\therefore \mathcal{L}^{-1}\left[\frac{3}{s(s-2)^2}\right] = \mathcal{L}^{-1}\left[\frac{0.75}{s} + \frac{1.5}{(s-2)^2} - \frac{0.75}{s-2}\right]$$

$$= \underline{0.75 + 1.5t\epsilon^{2t} - 0.75\epsilon^{2t}} \quad \text{or equivalent} \quad t \geq 0$$

19.5   Let's approach the series L-R-C circuit on a different tack. Here we will solve *simultaneously* for $i(t)$ and $v_c(t)$, $t \geq 0$.

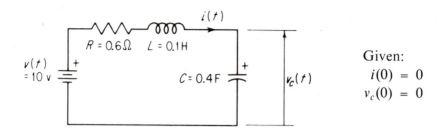

Given:

$i(0) = 0$

$v_c(0) = 0$

From Kirchhoff's voltage law

$$L \frac{di(t)}{dt} + Ri(t) + v_c(t) = v(t)$$

and for a capacitor, the $v - i$ relation is:

$$C \frac{dv_c(t)}{dt} = i(t)$$

Transforming, and rearranging slightly:

$$\{Ls + R\}I(s) + \quad V_c(s) = V(s) \qquad \text{(I.C. zero)}$$

$$-I(s) + Cs V_c(s) = 0 \qquad \text{(I.C. zero)}$$

Substituting the data:

$$\{0.1s + 0.6\}I(s) + \quad V_c(s) = 10/s$$

$$-I(s) + 0.4s V_c(s) = 0$$

Whence, using determinants (Cramer's rule):

$$I(s) =$$

Put your answer in the form: $\dfrac{\text{constant}}{(s^2 + \qquad s + \qquad)}$

$$I(s) = \frac{\begin{vmatrix} 10/s & 1 \\ 0 & 0.4s \end{vmatrix}}{\begin{vmatrix} 0.1s + 0.6 & 1 \\ -1 & 0.4s \end{vmatrix}} = \frac{4}{0.04(s^2 + 6s + 25)} = \frac{100}{s^2 + 6s + 25}$$

19.6   We have $I(s) = \dfrac{100}{s^2 + 6s + 25}$

Now, EXPAND $I(s)$ in P.F.'s and then FIND $i(t)$:

A FINAL SET OF PRACTICE PROBLEMS

In the next three sections you will be presented with three final problems which together cover the range of D.E.'s that we have considered.

Use your semi-notes, but *do not* look back at earlier sections of the program unless you get *completely* "stuck."

The first problem will require between 20 and 60 minutes of work, depending upon how many algebraic errors you make. It will pay to *be careful*!

$s^2 + 6s + 25 = 0$ has roots of $-3 + j4$ and $-3 - j4$.

Thus $\quad I(s) = \dfrac{100}{(s + 3 - j4)(s + 3 + j4)} = \dfrac{A}{s + 3 - j4} + \dfrac{B}{s + 3 + j4}$

(in the correct order to match line 11 of the semi-notes.)

$$A = \dfrac{100(\cancel{s + 3 - j4})}{(\cancel{s + 3 - j4})(s + 3 + j4)}\bigg|_{s = -3 + j4} = \dfrac{100}{j8} = -j12.5$$

In polar form (see diagram):

$\quad A = 12.5\,\epsilon^{-j90°}$

$\therefore B = 12.5\,\epsilon^{+j90°}$ $\quad$ (complex conjugate)

and, from line 11 of the semi-notes:

$\quad i(t) = 25\,\epsilon^{-3t}\cos(4t - 90°)$ amp $\quad t \geq 0$ $\quad$ (or equivalent)

Comment:

If we solved for $V_c(s)$ by determinants, we would obtain

$$V_c(s) = \dfrac{250}{s(s^2 + 6s + 25)}$$

and taking the inverse transform we would find that:

$\quad v_c(t) = 10 + 12.5\,\epsilon^{-3t}\cos(4t + 143°.2)$ $\quad t \geq 0$

Work this for yourself if you would like more practice.

SLIP THE MASK UNDER PAGE 171.

You may write on the page opposite if you need additional space.

20.1   SOLVE $\dfrac{dy}{dt} + 2y + 2 \displaystyle\int_0^t y\,dt - 2 = 2t$ with I.C. $y(0) = 0$:

Answer: $y(t) = $ _____ $t \geq 0$

WHEN YOU HAVE WRITTEN DOWN YOUR ANSWER, TURN THE PAGE.

20.2   Did you obtain the result

$$y(t) = 1 + 1.414\epsilon^{-t}\cos(t - 135°) = 1 + 1.414\epsilon^{-t}\cos(t + 225°) \qquad t \geq 0$$

or equivalent?   (*circle one*)

Comment:
Did you check your work at least to the extent of calculating $y(0)$ from your solution, for comparison with the I.C., $y(0) = 0$? You should have!

IF YOUR ANSWER WAS CORRECT, MOVE THE MASK AHEAD AND TURN THE PAGE.

IF YOUR ANSWER IS INCORRECT, GO TO THE NEXT FRAME, BELOW.
                                       DO *NOT* MOVE THE MASK DOWN.

20.3   Did you find that

$$Y(s) = \frac{2(s + 1)}{s(s^2 + 2s + 2)} = \frac{2(s + 1)}{s(s + 1 - j)(s + 1 + j)}$$

or other equivalent?

IF YOU AGREE WITH THIS $Y(s)$, CORRECT YOUR P.F. EXPANSION AND/
                          OR INVERSE TRANSFORMATION AND
                          THEN RETURN TO FRAME **20.2**.

IF YOUR $Y(s)$ IS INCORRECT, CORRECT YOUR TRANSFORMATION OF
                          THE ORIGINAL D.E. AND/OR SOLUTION
                          FOR $Y(s)$. RECOMPUTE $y(t)$, AND RETURN
                          TO FRAME **20.2**.

IF YOU ARE IN NEED OF HELP, CHECK THE BOX

AND LOOK UNDER THE MASK.

Given: $\dfrac{dy}{dt} + 2y + 2\displaystyle\int_{0}^{t} y\,dt - 2 = 2t$ with $y(0) = 0$

The transformed equation is:

$$\overset{\displaystyle 0}{sY(s) - y(\cancel{0})} + \underset{\text{transform of }2y(t)}{2Y(s)} + 2Y(s)/s - 2/s = \underset{\text{transform of }2t}{2/s^2}$$

Solving,

$$\{s + 2 + 2/s\}\,Y(s) = \frac{2}{s^2} + \frac{2}{s} = \frac{2(s + 1)}{s^2}$$

$$Y(s) = \frac{2(s + 1)}{s(s^2 + 2s + 2)} = \frac{2(s + 1)}{s(s + 1 - j)(s + 1 + j)}$$

$$= \frac{A}{s} + \frac{B}{s + 1 - j} + \frac{C}{s + 1 + j}$$

Evaluating A, B, and C we find (using the normal method)

$A = 1$

$B = 0.707\epsilon^{-j135°}$

$C = 0.707\epsilon^{+j135°}$

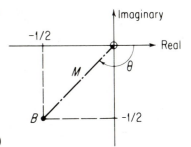

Therefore, $\underline{y(t) = 1 + 1.414\epsilon^{-t}\cos(t - 135°)}$ $t \geq 0$

or other equivalent.

The second of the three problems should take you 15 to 40 minutes to solve. Again, you may work on the page opposite if you need more space.

21.1   SOLVE $2\dfrac{d^2y}{dt^2} + 14\dfrac{dy}{dt} + 20y = 6\epsilon^{-2t}$ with I.C.'s $y(0) = 0$ and $\left.\dfrac{dy}{dt}\right|_0 = 3$

Answer: _____

WHEN YOU HAVE GIVEN YOUR ANSWER, TURN THE PAGE.

21.2   Did you obtain the result

$$y(t) = t\epsilon^{-2t} + 0.667\epsilon^{-2t} - 0.667\epsilon^{-5t} \qquad t \geq 0$$

yes/no ?

Comments:

1. The solution checks with the I.C. $y(0) = 0$.
2. If you differentiate the solution and put $t = 0$, the result checks with the I.C. $dy/dt \big|_0 = 3$.
3. Your answer is not disqualified if you neglected to write "$t \geq 0$," but remember that your solution holds *only* for $t \geq 0$.

IF YOUR ANSWER WAS CORRECT, MOVE THE MASK AHEAD AND TURN THE PAGE.

OTHERWISE, GO TO THE NEXT FRAME, BELOW. DO *NOT* MOVE THE MASK DOWN.

21.3   Did you find that

$$Y(s) = \frac{3(s + 3)}{(s^2 + 7s + 10)(s + 2)} = \frac{3(s + 3)}{(s + 2)^2(s + 5)}$$

yes/no ?

IF YOU OBTAINED THIS $Y(s)$, CORRECT YOUR P.F. EXPANSION AND/ OR INVERSE TRANSFORMATION AND THEN RETURN TO FRAME **21.2**.

IF YOUR $Y(s)$ IS INCORRECT, CORRECT YOUR TRANSFORMATION OF THE D.E. AND/OR SOLUTION FOR $Y(s)$. RECOMPUTE $y(t)$ AND RETURN TO FRAME **21.2**.

IF YOU NEED HELP, CHECK THE BOX

AND LOOK UNDER THE MASK.

Given: $2\dfrac{d^2y}{dt^2} + 14\dfrac{dy}{dt} + 20y = 6\epsilon^{-2t}$ with $y(0) = 0$ and $\dfrac{dy}{dt}\bigg|_0 = 3$

Thus $2\left\{s^2Y(s) - sy(0) - \dfrac{dy}{dt}\bigg|_0\right\} + 14\{sY(s) - y(0)\} + 20Y(s) = 6/(s + 2)$

(with the $sy(0)$, $\frac{dy}{dt}\big|_0$, and $y(0)$ terms marked as 0)

Solving, $Y(s) = \dfrac{3(s + 3)}{(s^2 + 7s + 10)(s + 2)} = \dfrac{3(s + 3)}{(s + 2)^2(s + 5)}$

$$= \dfrac{A}{(s + 2)^2} + \dfrac{B}{s + 2} + \dfrac{C}{s + 5}$$

Evaluating A, B, and C:

$A = 1$

$B = 0.667$

$C = -0.667$

Therefore $\underline{y(t) = t\epsilon^{-2t} + 0.667\epsilon^{-2t} - 0.667\epsilon^{-5t} \qquad t \geq 0}$

The next (and last) problem will take you 20 to 40 minutes to solve. Be careful with the algebra. You may work on the opposite page if you need the space.

22.1   *SOLVE* the simultaneous equations

$$\left.\begin{aligned} \frac{dy}{dt} - z &= -1 \\[2mm] -y + \frac{dz}{dt} &= t \end{aligned}\right\}$$ with the I.C.'s $y(0) = 0$ and $z(0) = -1$.

Answer: $y(t) =$ _____

$z(t) =$ _____

WHEN YOU HAVE OBTAINED YOUR ANSWER, TURN THE PAGE.

22.2   Do you agree with the result

$$y(t) = \underline{-t + \tfrac{1}{2}\epsilon^{-t} - \tfrac{1}{2}\epsilon^{t}} \qquad t \geq 0$$

$$z(t) = \underline{\quad -\tfrac{1}{2}\epsilon^{-t} - \tfrac{1}{2}\epsilon^{t}} \qquad t \geq 0$$

 ?

Comments:

1. If you found $y(t)$ and $z(t)$ without *any* error you were either *very* careful, or very lucky!
2. Did you remember the "$t \geq 0$" this time?
3. Did you check your solution against the I.C.'s $y(0) = 0$ and $z(0) = -1$?

IF YOU AGREED WITH THE ANSWER, ABOVE, REMOVE THE MASK AND TURN THE PAGE.

IF YOU DID NOT AGREE, GO TO THE NEXT FRAME, BELOW.   DO *NOT* MOVE THE MASK.

22.3   Did you find that

$$Y(s) = \frac{-2s^2 + 1}{s^2(s^2 - 1)}$$

and $$Z(s) = \frac{-s}{s^2 - 1}$$

 ?

IF "YES," CORRECT YOUR P.F. EXPANSION AND/OR INVERSE TRANS-
    FORMATION AND THEN RETURN TO FRAME **22.2**.

IF "NO," CORRECT YOUR TRANSFORMATION OF THE D.E.'s AND/OR
    YOUR SOLUTION (USING DETERMINANTS) OF THE TRANS-
    FORMED EQUATIONS.   THEN RECOMPUTE $y(t)$ AND $z(t)$ AND
    RETURN TO FRAME **22.2**.

IF YOU NEED HELP, CHECK THE BOX

AND LOOK UNDER THE MASK.

Given
$$\frac{dy}{dt} - z = -1$$

$$-y + \frac{dz}{dt} = t$$

with I.C.'s $y(0) = 0$ and $z(0) = -1$

Transforming and rearranging,

$$sY(s) - Z(s) = -1/s$$

$$-Y(s) + sZ(s) = \frac{1}{s^2} - 1 = -\frac{s^2 - 1}{s^2}$$

Solving,

$$Y(s) = \frac{\begin{vmatrix} -1/s & -1 \\ -(s^2 - 1)/s^2 & s \end{vmatrix}}{\begin{vmatrix} s & -1 \\ -1 & s \end{vmatrix}} = \frac{-1 - \dfrac{s^2 - 1}{s^2}}{s^2 - 1} = \frac{-2s^2 + 1}{s^2(s^2 - 1)}$$

$$= \frac{A}{s^2} + \frac{B}{s} + \frac{C}{s + 1} + \frac{D}{s - 1}$$

Whence, evaluating A, B, C, and D, we find $\underline{y(t) = -t + \frac{1}{2}\epsilon^{-t} - \frac{1}{2}\epsilon^{t}} \qquad \underline{t \geq 0}$

Similarly, if we solve for $Z(s)$ and expand, $\underline{z(t) = \qquad - \frac{1}{2}\epsilon^{-t} - \frac{1}{2}\epsilon^{t}} \qquad \underline{t \geq 0}$

YOU WILL NO LONGER NEED THE MASK: A SET OF COMPLETED SEMI-
NOTES WILL BE FOUND INSIDE THE BACK COVER.

SOME FINAL COMMENTS

You have now developed a skill in the Laplace Transform solution of differential equations which will be the basis for a large part of your continuing work in systems analysis,
as well as in other subjects where differential equations are used.

If now, or later, you wish to review the subject matter of this program, the TABLE OF
CONTENTS will be a helpful guide.

IF YOU WISH TO TEST YOURSELF, CONTINUE TO PAGE 185.

183

| A SAMPLE EXAMINATION |

Working time: 2 hours

If you are going to test yourself seriously, work the following exam in the allotted time of 2 hours. Do *not* refer to *anything* other than the completed semi-notes inside the back cover.

To help you allocate your time, estimated working times have been noted next to each question. These add to 120 minutes. Each question is worth as many points as the given working time. The highest possible score is thus 120.

Don't be dismayed by the number of "theory" questions. The applications *are* more important, but we wanted to give you an opportunity to test yourself on *all* the topics covered by the program.

185

1. Using the *definition* of the Laplace Transform (see back cover), FIND the L.T. of the constant, A:

(5 min)

---

2. Using the definition of the L.T. and the fact that $\int u \, dv = uv - \int v \, du$, FIND $\mathcal{L}[df(t)/dt]$:

(10 min)

---

3. Using the table of transforms, and the fact that $\cosh \beta t = \frac{1}{2}\{\epsilon^{\beta t} + \epsilon^{-\beta t}\}$, FIND

$\mathcal{L}[\cosh \beta t]$:

(5 min)

---

4. Given the transforms of the first and second derivatives (inside the back cover), DERIVE the result

$$\mathcal{L}[d^3f/dt^3] = s^3 F(s) - s^2 f(0) - s \left.\frac{df}{dt}\right|_0 - \left.\frac{d^2f}{dt^2}\right|_0$$

(10 min)

---

5.   DERIVE Theorem 5 *from* Theorem 2.
(*Hint:* Define the auxiliary function $g(t)$ by the relation $f(t) = dg/dt$.)

(10 min)

---

6.   FIND $\mathcal{L}[3\epsilon^{-4t}\sin 5t]$: (no special algebraic form is required for the answer.)

(5 min)

---

7.   Using the fact that $\mathcal{L}[t\epsilon^{at}] = \dfrac{1}{(s-a)^2}$, FIND $\mathcal{L}\left[\displaystyle\int_0^t t\epsilon^{at}\,dt\right]$:

(5 min)

8. FIND the inverse transform of $\dfrac{s + 1}{s^2 + 2s}$:

(5 min)

---

9. FIND $\mathcal{L}^{-1}\left[\dfrac{1}{s(s^2 - 2s + 5)}\right]$, putting your answer in real form:

(10 min)

189

10. FIND $\mathcal{L}^{-1}\left[\dfrac{1}{(s+1)(s+2)^2}\right]$:

(10 min)

11. SOLVE the equation $\dfrac{d^2x}{dt^2} - 3\dfrac{dx}{dt} + 2x = \epsilon^{3t}$ given that $x(0) = 1$, and $dx/dt\,|_0 = 1$:

(20 min)

12. SOLVE the equations

$$3\frac{dx}{dt} + 2x + \frac{dy}{dt} = 1$$
$$\frac{dx}{dt} + 4\frac{dy}{dt} + 3y = 0$$

for the unknown, $y(t)$, given that $x(0) = y(0) = 0$:

(25 min)

The test solution follows. You can grade yourself if you wish, making some estimate of partial credit where you have erred.

#	Value	Score
1	5	
2	10	
3	5	
4	10	
5	10	
6	5	
7	5	
8	5	
9	10	
10	10	
11	20	
12	25	
Total	120	

Answers to sample examination questions:

1. $\displaystyle\int_0^\infty A\epsilon^{-st}\,dt = \frac{A}{s}$ (assuming $s > 0$ or $\Re e(s) > 0$)

2. See frames **2.1-2.3**

3. $\dfrac{s}{s^2 - \beta^2}$ (or equivalent)

4. a) Let $g(t) = \dfrac{df}{dt}$

$$G(s) = sF(s) - f(0)$$

$$\mathcal{L}\left[\frac{d^3f}{dt^3}\right] = \mathcal{L}\left[\frac{d^2g}{dt^2}\right]$$

$$= s^2 G(s) - sg(0) - \left.\frac{dg}{dt}\right|_0$$

$$= s^2\{sF(s) - f(0)\} - s\left.\frac{df}{dt}\right|_0 - \left.\frac{d^2f}{dt^2}\right|_0$$

or   b) Let $g(t) = \dfrac{d^2f}{dt^2}$

$$G(s) = s^2 F(s) - sf(0) - \left.\frac{df}{dt}\right|_0$$

$$\mathcal{L}\left[\frac{d^3f}{dt^3}\right] = \mathcal{L}\left[\frac{dg}{dt}\right]$$

$$= sG(s) - g(0)$$

$$= s\left\{s^2 F(s) - sf(0) - \left.\frac{df}{dt}\right|_0\right\} - \left.\frac{d^2f}{dt^2}\right|_0$$

5. See frames **8.1–8.3**

6. $\dfrac{15}{s^2 + 8s + 41}$ (Theorem 6)

7. $\dfrac{1}{s(s-a)^2}$

8. $\dfrac{1}{2} + \dfrac{1}{2}\,\epsilon^{-2t}$ $t \geq 0$

9. $\dfrac{1}{s(s^2 - 2s + 5)} = \dfrac{0.2}{s} + \dfrac{0.112\epsilon^{-j153^\circ.4}}{s - 1 - j2} + \dfrac{0.112\epsilon^{j153^\circ.4}}{s - 1 + j2}$

and $\mathcal{L}^{-1}\left[\dfrac{1}{s(s^2 - 2s + 5)}\right] = 0.2 + 0.224\epsilon^t \cos\left(2t - 153^\circ.4\right)$ $t \geq 0$

10. $\dfrac{1}{(s+1)(s+2)^2} = \dfrac{1}{s+1} - \dfrac{1}{(s+2)^2} - \dfrac{1}{s+2}$

and $\mathcal{L}^{-1}\left[\dfrac{1}{(s+1)(s+2)^2}\right] = \epsilon^{-t} - t\epsilon^{-2t} - \epsilon^{-2t}$ $t \geq 0$

11. $X(s) = \dfrac{s^2 - 5s + 7}{(s^2 - 3s + 2)(s - 3)} = \dfrac{3/2}{s-1} - \dfrac{1}{s-2} + \dfrac{1/2}{s-3}$

and $x(t) = \dfrac{3}{2}e^t - \epsilon^{2t} + \dfrac{1}{2}\epsilon^{3t}$ $t \geq 0$

12. $Y(s) = \dfrac{-1}{11s^2 + 17s + 6} = \dfrac{-\dfrac{1}{11}}{s^2 + \dfrac{17}{11}s + \dfrac{6}{11}} = \dfrac{-\dfrac{1}{5}}{s + \dfrac{6}{11}} + \dfrac{\dfrac{1}{5}}{s + 1}$

and $y(t) = \dfrac{1}{5}\left(\epsilon^{-t} - \epsilon^{-6t/11}\right)$ $t \geq 0$

$$\boxed{\textbf{BIBLIOGRAPHY}}$$

If you want to see more examples and applications, or if you wish to investigate some of the more advanced aspects of the L.T. method, the following references will be helpful.

1. H. S. Carslaw and J. C. Jaeger, "*Operational Methods in Applied Mathematics*," Oxford University Press, London, 1941.

    One of the earliest texts on the subject, and easy to read. Uses "*p*" instead of "*s*."

2. R. V. Churchill, "*Operational Mathematics*," 2nd Edition, McGraw-Hill, New York, N.Y., 1958.

    A more rigorous approach to operational methods, of which the L.T. is one.

3. M. F. Gardner and J. L. Barnes, "*Transients in Linear Systems*," John Wiley & Sons, New York, N.Y., 1942.

    The first text emphasizing the applications of the L.T. method to dynamic system analysis.

4. M. E. Van Valkenburg, "*Network Analysis*," 2nd Ed., Prentice-Hall, Inc., Englewood Cliffs, N.J., 1964.

    Develops the concepts of circuit analysis from a foundation based on the L.T.